A TOPICAL DICTIONARY OF STATISTICS

Gary L. Tietjen

A TOPICAL DICTIONARY OF STATISTICS

Gary L. Tietjen

CHAPMAN AND HALL
New York   London

First published 1986
by Chapman and Hall
29 West 35 Street, New York, N.Y. 10001

Published in Great Britain by
Chapman and Hall Ltd
11 New Fetter Lane, London EC4P 4EE

©1986 Chapman and Hall

Printed in the United States of America

Library of Congress Cataloging-in-Publication Data
Tietjen, Gary L.
  A topical dictionary of statistics.

  Bibliography: p.
  Includes index.
  1. Mathematical statistics—Terminology. I. Title.
QA276.14.T54 1986   519.5'03'21   86-11716
ISBN 0-412-01201-4

Contents

Preface

Statistics is the accepted body of methods for summarizing or describing data and drawing conclusions from the summary measures. Everyone who has data to summarize thus needs some knowledge of statistics. The first step in gaining that knowledge is to master the professional jargon. This dictionary is geared to offer more than the usual string of isolated and independent definitions: it provides also the context, applications, and related terminology.

The intended audience falls into five groups with rather different needs: (1) professional statisticians who need to recall a definition, (2) scientists in disciplines other than statistics who need to know the acceptable methods of summarizing data, (3) students of statistics who need to broaden their knowledge of their subject matter and make constant reference to it, (4) managers who will be reading statistical reports written by their employees, and (5) journalists who need to interpret government or scientific reports and transmit the information to the public.

In every case the word or phrase to be defined should be looked up in the alphabetical index, which will refer the reader to a page in the text. The professional statistician may then find the word and its definition and be finished. Other readers are no doubt looking for *more* information—for background, related words, and an understanding of how this topic fits into the scheme of things. For this purpose the dictionary has been arranged topically rather than alphabetically, and in connected discourse rather than in paragraphs related only by the first few letters of one word. Depending on how much knowledge of the subject is desired, the reader will want to go back a few paragraphs or to the beginning of the chapter and read further. I have even been assured by some professors of statistics that students of statistics will benefit by reading the dictionary from cover to cover. That will give some idea of the "big picture" and help substitute for gaps in educational training.

The first chapter, *Summarizing Data*, has been written particularly for those who have no background in statistics and may be worth their while as an introduction to the book.

In trying to meet the needs of so many, I have tried to stay one level below what a theoretical statistician would like to see. In doing so, I run the risk of falling a little short of absolute rigor, but I have tried not to sacrifice the things that matter in applications. I hope this effort will be worthwhile for the manager and journalist. The question of what to include and what to leave out has arisen many times. There are many specialties in statistics and I have decided to omit terms which are peculiar to a small audience. On the other hand, I have included a number of general but rarely used terms for the sake of completeness, and I can only hope that the reader will bear with me on this matter. There are terms whose use is discouraged and areas where one needs to take great care to avoid misstatements. I have tried to point these out to the reader.

The nonmathematical reader need not be discouraged by the mathematical symbols which are used frequently. They are really a rather simple shorthand.

The symbols appearing most often are the following: (1) $\int_a^b f(x)dx$ which is read "the integral of $f(x)$ from a to b" and means the area under the curve $f(x)$ and between the values of $x = a$ and $x = b$ on the x-axis. if the limits are infinite, i.e. if a is $-\infty$ and b is ∞, the area is taken over the entire x-axis (2) the Greek letter Σ means "the sum of" so that $\sum_{i=1}^{N} x_i$ means "the sum of the x_i from $i = 1$ to $i = N$", i.e. $x_1 + x_2 + \cdots + x_N$. For brevity, the subscripts are sometimes omitted when the meaning is clear. Thus the symbol Σx means Σx_i, where i goes from 1 to N unless otherwise indicated. Products are denoted with a Π in place of Σ. The symbol $x!$ means the product of the positive integers from 1 to x, i.e., $5! = 1 \times 2 \times 3 \times 4 \times 5$. The binomial symbol $\begin{bmatrix} N \\ n \end{bmatrix}$ is equal to $N!/(N - n)!n!$ and is the number of ways of selecting a sample of n items from a set of N items (4) the symbol df/dx (read "the derivative of f with respect to x") refers to the slope of the line tangent to the curve $f(x)$ at a point x. (For a given change in x, the slope of a line is the change in y divided by the change in x). If f is a function of more than one variable, $\partial f/\partial x$ is the derivative of f with respect to x while the other variables are held constant (it is read "the partial of f with respect to x").

I have followed the convention of using capital letters for random variables and the corresponding lower case letters for their realizations (or for non-random variables). Greek letters denote parameters of a distribution (for a particular distribution, a parameter is a constant, but it varies from one distribution to another). Parameters are usually unknown constants which have to be estimated, and their estimates are denoted by the same Greek letter with a carat (^) or a tilde (˜) placed above it. A tilde in the middle of the line, however, is read "is distributed as". The symbol $E(X)$ is read "the expected value of X" and is an average of the values taken on by the random variable X. The symbol $\sqrt{x}$ is the positive square root of x and is equivalent to $x^{1/2}$. The use of $\exp(x)$ for e^x is purely for typesetting convenience. The natural logarithm of x (log base e, where e is a constant approximately equal to 2.718) is denoted by $ln(x)$.

This dictionary is by no means a solo production. At one time or another I pressed nearly everyone in the statistics group at Los Alamos into providing some definitions or reading those I had written. Consultants to the group received the same treatment. Dennis Cook generously wrote the major part of the chapter on regression, while George Milliken wrote much of the chapter on Experimental Design. I benefitted especially from a critical review of the first draft by Jay Conover. Ben Duran followed with many valuable suggestions. My good friend S. Juan lent constant support and encouragement. Kay Grady and Corinne Ortiz very competently typed the manuscript and endured revision after revision without complaint. Finally, I am most grateful to my wife who took care of many other duties while I wrote.

Gary Tietjen
Los Alamos
New Mexico
May, 1986

A TOPICAL DICTIONARY OF STATISTICS

Gary L. Tietjen

1

Summarizing Data

This chapter is intended to introduce the basic ideas of statistics to the layman. Statistics is the accepted method of summarizing or describing data and then drawing inferences from the summary measures. Suppose, for example, that a company has made a process change in manufacturing its light bulbs and hopes that the new bulb (Type B) will have a longer lifetime than the old (Type A). Having experimented previously, the company knows that even though the bulbs are treated identically they will vary considerably (60 to 90 hours) in the length of time they will last. That variability, which is a property of almost all manufactured products, is called *inherent variability*. Since we cannot predict how long a bulb will burn, we describe its lifetime as a *random variable*. The company does know that the largest fraction of the bulbs burn about 75 hours, that those with lives of 70 and 80 hours are about equally frequent (but less common than lifetimes of 75 hours), and that those with lives of 65 and 85 hours are even less frequent.

For bulb A we can think of all the past production as a *population* of bulbs with lifetimes that we denote by x. For bulb B the population is mostly conceptual; it consists of bulbs that *will* be produced by the new process. We let y denote the lifetime of a bulb of Type B.

1

The aim of management in this instance is to evaluate the performance of the Type B bulbs and to compare it with that of Type A. The first thing to do is to picture the situation. It is obviously neither possible nor desirable to test the lifetimes of all the bulbs. Fortunately the company has tested 100 Type A bulbs in the past. The readings range from 60 hours to nearly 100 hours burning time. That portion of the population will ordinarily be called a sample, but the word implies that there are some restrictions in the way the bulbs to be tested are selected. When there are no restrictions, we shall refer to the portion as a *batch*. (In this chapter we are adopting some of the terminology coined recently by John Tukey for the set of techniques that he calls *Exploratory Data Analysis* [EDA]. His terminology, while not yet standard, has come into rather wide usage. Exploratory data analysis is a first look—a quick glance—at the data and is usually followed by a *confirmatory data analysis*, using the techniques of classical statistics.)

A time-honored method of graphically portraying the data in the batch of 100 units is to construct a *histogram* of the data. This is done by dividing the interval of possible lifetimes (60 to 100 hours) into k subintervals of equal width called *class intervals*. There is no prescribed way of deciding how many intervals to use, but perhaps 10 will do here: 60–65, 65–70, . . . We then count the number of units (n_1, n_2, . . .) that fall into each interval. We next draw a series of adjacent rectangles, using the class intervals as widths and the frequencies n_1, n_2, . . . , n_k as heights. The histogram shows the *distribution of frequencies* of the lifetimes. The graph can be improved somewhat by using *relative frequencies* (n_1/T, n_2/T, . . . n_k/T, where $T = n_1 + n_2 + \ldots + n_k$) as the heights of the rectangles. Much can be inferred from the histogram. We can decide, for example, what percentage of the bulbs in the sample have lifetimes of less than 70 hours, what percentage burn between 80 and 90 hours, etc. If we took larger and larger samples, we could make the class intervals narrower and narrower until the tops approached a smooth, continuous curve. If such a curve were drawn across the midpoints of the tops of the rectangles, it would represent the *frequency distribution* or *probability density function* (pdf) for the population. We see that the relative frequencies sum to 1; hence it is not hard to believe that the area under the frequency distribution is 1. Further, the area under the curve and within an interval (a,b) of lifetimes is approximately the relative frequency of lifetimes within the interval. The area between a and b is, in fact, the limiting value of the relative frequency and is called the *probability* that the lifetime is between a and b. We thus see that a random variable has a distribution of probabilities associated with it.

The EDA version of a histogram may be quicker to construct and is called a *stem-and-leaf plot*. It will be lying on its side. The stems replace class intervals and in this case would be the first digit of the lifetime. The second digit constitutes the leaves. By tallying the leaves to the right of their stem as we come to them in the data set, we get the stem and leaf plot of the data (provided that we give the same width to each leaf). If we decide that we do not have enough stems, a period following the stem can represent a stem with leaves 0–4, while an asterisk following the stem accompanies leaves 5–9. Three-digit numbers can be represented by dropping the last digit or by using 2-digit stems.

There is a way of abbreviating a histogram even further. Let us first *rank* the n data points in ascending order so that the smallest point has rank 1, the second-smallest rank 2, and so on to the largest, which has rank n. Now we rank the data in descending order so that the smallest point has rank n and the largest rank 1. The *depth* of a data point is the minimum of the 2 ranks the data point can have. The largest and smallest points, called *extremes*, have depth 1. The middle observation or *median* has depth $(1 + n)/2$. When the depth is not an integer, we average the 2 data points with depths on either side of the indicated 1. Thus, if there is an even number of points, the median is the average of the 2 middle points. The *hinges* are halfway between the extremes and the median; they are the points with depth $(1 + m)/2$, where m is the integer part of the depth of the median. Similarly the *eighths* are points with depth $(1 + h)/2$, where h is the integer part of the depth of the hinges.

A rather neat summary of the histogram is made by plotting the extremes, hinges, and median on a vertical line. A "long, thinnish box" (about ⅛ inch wide) is drawn so that the hinges are at the top and bottom of the box. A horizontal line through the box marks the location of the median. A vertical line connects the extremes with the hinges. That 5-point summary is called a *box-and-whisker plot*. The middle 50 percent of the data lie inside the box; the lower 25 percent of the data are in the lower whisker and the upper 25 percent in the upper whisker. The lower half of the data are below the median and the other half above it.

Another useful plot, very similar to a box-and-whisker plot, is a *schematic plot*. The *H-spread* is the distance between the hinges. A *step* is 1.5 times that distance. An *inner fence* is 1 step beyond the hinges, and an *outer fence* is 2 steps beyond the hinges. The data point closest to the inner fence but inside of it is an *adjacent* point. *Outside* values are those between the inner and outer fences, while those beyond the outer fence are *far out* points. The box for the schematic plot is constructed as before, but the whiskers are dashed and extended only to the adjacent values and end with a short dashed horizontal line. The outside values are labeled separately and the far out values labeled "impressively."

Let us return to our example. A batch of bulbs is taken from the production line and tested. Either box-and-whisker plots or schematic plots are constructed so that the plots for the 2 types of bulbs parallel each other. A visual comparison of the 2 sets of data can now be made. Symmetry of the histogram (with the median near the middle of the box and the whiskers of about equal length) is a good framework for comparing the medians. More important than symmetry is an approximate equality in spread, as shown by box length and whisker length. If either of those conditions is violated seriously, we will want to *re-express* the data (*transform* the data is classical terminology) before comparing the medians. The schematic plot is preferred to the box-and-whisker plot if there are several outside or far out points. If the 2 distributions are moderately symmetrical and close in spread as judged by the eye, the medians will tell about how far apart the average lifetimes will be. Sometimes the *trimeans* (sum of hinges plus twice the median, all divided by 4) are used as an estimate of the "center" of the distribution.

When strong asymmetry/inequality of spread is present, the re-expression is ordinarily done by transforming the data to logs of the data. If that does not work, a square root transformation is made. Negative reciprocals are the third choice. All of those choices preserve the ranks and depths of the data points. Tukey has suggested a *ladder of transformations* . . . X^3, X^2, X, log X, $-1/X$, $-1/X^2$, $-1/X^3$, . . . , where the transformation most likely to help is chosen by plotting the log H-spread (y) against the log of the median (x) for the several populations being compared. If the slope of a line drawn by eye is close to ½, the square root should help. If the slope is somewhat larger than ½, logarithms will be more likely to be useful. In other words, choose the re-expression more apt to result in a horizontal line through the transformed points.

We now give some thought to a confirmatory or classical approach, which might follow the quick EDA look at the data. The EDA approach should have given us a rather good "feel" of the data and any unusual structure that might be present. From that we may have reached some preliminary conclusions. At other times the differences between the 2 bulbs may have been so obvious that no statistics seem to be needed. Regardless, the investigator needs numbers to put in the report. Just how large *are* the differences between medians? How significant are the results? That is an area for classical statistics.

A great many measurements in nature are approximately *normally distributed* (for a good reason to be discussed later). That means that the *probability density function* (the frequency distribution) is symmetrical and bell-shaped. That distribution is so familiar and so ubiquitous that in many cases the statistician just assumes that the measurements are normal (in most cases it does not matter too much if that assumption is slightly off base). In situations where the assumption seems to be badly off, the statistician may test it with a *goodness-of-fit test*. The "center" of the bell, the place where it has a "peak," is called the *mean* and designated by the greek letter µ. The distance from

μ to the point at which the curvature changes from downward to upward is the *standard deviation*, designed by σ. The area under the curve is 1. The area in the interval $\mu \pm \sigma$ is about 68 percent of the total; the area between $\mu - 2\sigma$ and $\mu + 2\sigma$ is about 95 percent of the total, and the area between $\mu - 3\sigma$ and $\mu + 3\sigma$ is well over 99 percent of the total. The curve is completely characterized by a knowledge of μ and σ.

Returning to the bulbs and assuming a normal distribution for each of the 2 populations, we can condense or summarize the data even further. What single number is typical of or characterizes or summarizes the lifetimes in the sample? The *average* lifetime of the bulbs immediately comes to mind, but what shall we average? We think of the entire past production of bulb A as the *population* of interest and let x_i be the lifetime of the i-th bulb ($i = 1, 2, \ldots N$) produced. The *population average* μ_x would then be the sum of the lifetimes divided by the number of bulbs in the sum $\mu_x = \Sigma x_i / N$. The number N, however, is in the millions, and there was no possible way to have gotten the necessary measurements. From the 100 measurements taken on the Type A bulbs, the *sample average* is $\bar{x} = \Sigma x_i / n$, where $n = 100$. The sample average is an *estimate* of the population average. The formula $\Sigma x_i / n$ is an *estimator* of μ_x. Having obtained $\bar{x}$, we would like to do the same thing for bulb B, which is just getting into production. How large a sample shall we take? With bulb A there was little choice: We took all the information available at the time. It seems intuitive that the larger the sample size the better the estimate (a property that will later be called *consistency*). We thus take the largest sample—say, m bulbs—we can afford; the cost will also involve the time spent in testing. The sample average is $\bar{y} = \Sigma y_i / m$. We could now compare $\bar{x}$ and $\bar{y}$ to see which is larger, but we do not have, as yet, a good standard with which to judge the difference. If the sample size m were small and if the difference between $\bar{x}$ and $\bar{y}$ were small, a different sample of bulbs might have yielded an average that would have reversed the order of $\bar{x}$ and $\bar{y}$. Whether normality holds or not, a useful measure of the scatter of the data around the sample mean is the *sample variance* $s_y^2 = \Sigma (y_i - \bar{y})^2 / (n-1)$, which estimates or approximates the population variance $\sigma_y^2 = \Sigma (y_i - \mu_y)^2 / N$, the average squared deviation from the mean. We divide the sample variance by $(n-1)$ instead of n because it can be shown that the average value of s_y^2 (with a divisor of n) is $(n-1)\sigma_y^2 / N$ σ_y^2. In other words, s_y^2 would be a *biased estimator* of σ_y^2, and the divisor $(n-1)$ is chosen to unbias it.

We can now express our uncertainty in the location of the sample mean μ_y by making an *interval estimate* of μ_y, which has a "high probability" of containing μ_y in the following sense: If a large number of such intervals were constructed from different samples, 95 percent of them would contain the population mean. That interval is called a 95 percent *confidence interval* for μ_y, as opposed to the *point estimate* $\bar{y}$ of μ_y. The interval is $\bar{y} \pm t\, s_y/\sqrt{n}$, where t is a number that depends upon the sample size n and is taken from tabled values of the Student's t distribution, a connection we will explore later.

Finally, we can test whether the population means for the 2 types of bulbs differ significantly, our hypothesis being that $\mu_y = \mu_x$. The sample means clearly differ only if the difference between them is larger than the variability within the measurements that make up the means. The difference $\bar{x} - \bar{y}$ is distributed with mean zero and standard deviation of $(s_y^2/m + s_x^2/n)^{1/2}$ if and only if $\mu_x = \mu_y$. Differences of about 2 of those standard deviations are not so unusual, but differences much larger than that are rare—so rare that we are willing to take a small risk (of the order of 5 percent) of being wrong and declare that μ_x is not equal to μ_y. In other words, we decide that one mean is greater than the other. The actual number of standard deviations by which $\bar{x}$ and $\bar{y}$ can differ without differing significantly is again found in the tables of the Student's t distribution and depends on n, m, and the size of the small risk we take of being wrong. It is one example of *hypothesis-testing*, a very valuable tool.

In this chapter we have touched on random variables and their probability distributions. We have seen how a histogram approximates the probability density function. Those matters are covered in Chapter 2. Lifetimes of electrical components frequently have distributions other than normal. A guide to other distributions and their uses is found in Chapter 3. We have examined estimators and estimates of the mean and variance of the normal distribution, and we have given some thought to the desirable properties of an estimator (consistency and unbiasedness). We have touched on the topic of hypothesis testing. The areas of estimation and hypothesis testing are taken up in detail in Chapter 4.

Suppose now that a special coating of the filament was the design change that resulted in the Type B bulbs. Some thought has been given to whether the life of the bulb might increase directly with the thickness of that coating. Some experiments are carried out using various thicknesses of the coating and testing of the life of each bulb. The data are plotted as lifetime (y) versus thickness (x), and it appears there is a linear relationship. The plot of the data points is called a *scattergram* or *scatterplot*. How to fit a straight line to the data is a problem in estimating the parameters (the slope and intercept) of a straight line. Again, hypothesis testing is used to decide whether the slope is zero (no change in lifetime with thickness) or not. Those matters form the content of Chapter 5 on *Regression*.

It may be that 3 different coatings can be applied, each differing in its composition. In order to test which of the 3 gives the longest average lifetime, we would design an experiment in which we would measure the lifetime of k bulbs with each coating. The analysis of those data would involve the *Analysis of Variance*, a subject taken up in Chapter 6: The Design of Experiments and Analysis of Variance.

The probability that a bulb will perform its function (burn) for a given length of time under given circumstances is the *reliability* of the bulb. The whole area of estimation and testing of lifetime data is treated in Chapter 7: Reliability and Survival Analysis.

It may have been of interest to estimate the intensity of light from the new bulbs as a function of the age of the bulb. If we had a continuous record of the intensity (or a test every 30 minutes, say) with time, we would have a *time series* and the analysis of the data would come under the chapter on Time Series and its parent: Stochastic Processes.

In the production of the new bulbs, it may be desirable or necessary for the manufacturer to assure himself continually that the thickness of the coating is uniform. To do that, he may check the thickness of 3 bulbs from each day's production. The techniques for obtaining such assurance are given in the chapter on *Quality Control*.

If we have 2 characteristics of interest, say lifetime and intensity, then there are 2 random variables to be considered simultaneously. That is a problem in *Multivariate Analysis*, which falls under the chapter of that name.

REFERENCES

For the EDA techniques in this chapter see Tukey, J. W. 1977. *Exploratory Data Analysis*. Reading, Mass: Addison-Wesley. The classical techniques will be explained later in detail.

2

Random Variables and Probability Distributions

When the average citizen sees a game of dice, he knows intuitively that the outcome of any 1 roll of the dice is unpredictable—that he is faced with a "chance" or "random" phenomenon. He sees quickly that there are 36 possible outcomes (for each of the 6 sides of die #1 he can get any of the 6 sides of die #2). He nevertheless realizes the possibilities of betting on the outcome when he sees that there is only 1 outcome, (1,1), which gives a "2," while 6 outcomes, (1,6), (2,5), (3,4), (4,3), (5,2), (6,1), give a "7." Thus the "probability" of a 2 is 1/36 and that of a 7 is 6/36. The set of outcomes or "scores" with their associated probabilities constitutes a *probability distribution* and is his best aid to intelligent betting. In that case the dice were treated or "shaken" alike. Individual outcomes differ, but in the "long-run" one can predict how often each outcome will occur. What the layman may not realize is that even in a very precise chemical experiment the outcome is random. The possible outcomes may lie within a narrow range, but when seemingly identical units are treated as much alike as possible, they still respond differently; there is still some "variability" in the outcome, and the chemist, with the aid of statistics, summarizes the data by telling his readers what they can bet on. We shall now repeat those ideas with more detail.

In an *experiment* the investigator observes the response to a given set of conditions. In some experiments the response is invariably the same, and we say that there is a *deterministic regularity* in the outcome. In other experiments, such as the toss of a die, the outcome is unpredictable, but the experiment has the next best property: The set of outcomes is known, and each outcome occurs with a certain relative frequency. Those *random experiments* (or *random trials* or *random events*), as they are called, are thus said to have a *statistical regularity* in the outcome. The relative frequency with which each outcome occurs approaches a stable limit, called the *probability* of that random event.

The set of all possible outcomes of a random experiment is called the *sample space* or *outcome space S*, and each outcome is a *sample point* ω in that space. An *event* is a subset of the sample space, but there may be some subsets that are not events. An event consisting of a single point is an *elementary event*. An event E is said to *occur* if the outcome ω is in E. In tossing a pair of dice, let $\omega = (2,3)$ be the outcome of a 2 on the first die and a 3 on the second. The point $(2,3)$ is an elementary event. If E is the event that the total shown on the 2 dice equals 5, E occurs if the outcome is $(1,4)$, $(4,1)$, $(2,3)$, or $(3,2)$.

The outcome of a coin-tossing experiment may be "heads" or "tails." In drawing a colored ball from an urn, the outcome may be "blue." In drawing a man from a group of men, the outcome might be Don or Joe. For the sake of a mathematical treatment (rather than a verbal one) we need to assign a real number to every outcome. The number assigned will depend upon our purpose. We might assign the number 1 to "heads" and 0 to "tails," which is useful if we are counting heads. We could assign 1 to "blue" and 0 to any other color; to each man we could assign his height in inches. Given a set A of "objects," a rule that assigns to each object in A 1 (and only 1) member of a set B is called a *function* with *domain A* and *range B*. A *random variable* is a function in which A is the set of outcomes and B consists of real numbers, including $\pm \infty$. In tossing a pair of die, we assign to each outcome the sum of the number of spots on the upward faces. It is important that the function representing the random variable be single-valued and real-valued. If the outcome is a number x in the interval $(-10, 10)$, say, we cannot let the square root of the outcome be the random variable for 2 reasons: (1) If $x = 4$, both $+2$ and -2 are square roots, and there must be only 1 number assignable to an outcome and (2) if $x = -4$, the square root is a complex number, and that is not allowed.

Associated with each random variable is a *set function* (so named because the domain consists of all possible events, each of which is a set). The set function assigns to each event the probability that the event will occur. Since probabilities are real numbers between zero and 1, the range of that set function is the interval [0,1]. For an event consisting of nonoverlapping subsets, the probability will be the sum of the probabilities associated with the subsets. The probability of the outcome space S, which is the union of all events or the "certain event," is 1.

For a random variable that assigns to each man his height in inches, say, the events of interest will usually be intervals or a union of intervals or the complement of an interval or an intersection of intervals. Examples are (a) the event that the man is between 5 and 6 feet tall, (b) the event that the man is not 5 to 6 feet tall, (c) the event that the man's height is a multiple of 12 inches. To assure ourselves that all sets of interest are events, we also include as events all the countable unions and intersections of events. The nonmathematical reader may now skip to the next paragraph. For the sake of rigor, we now restate that notion in more technical language by letting G be the collection of events on a sample space S. G is called a *field* (or *algebra*) of events if the complement of an event in G is an event in G and if the union and intersection of any 2 events in G is an event in G. G is a *σ-field* (or *σ-algebra*) if G also includes any countable union or countable intersection of events in G. A *probability measure* is a set function P that assigns to each event E in a field G a real number, $P(E)$, which is the probability that event E occurs, with 3 properties: (1) $0 \leqslant P(E) \leqslant 1$ (2) $P(S) = 1$ (3) if $\cup A_i$ is a countable union of disjoint (mutually exclusive) events in G, $P(\cup A_i) = \Sigma P(A_i)$. A probability measure is thus a triplet (S, G, P). Since the smallest σ-field that includes the intervals of interest to us is the σ-field of *Borel sets* B, it is natural to wish to extend our probability measure from G to B while keeping the properties P had on G. It is a standard result in measure theory that that can be done. We further remark that for complete rigor the random variable X is required to be a *Borel-measurable function* (so that we can always integrate) and that the integrals we use will be Lebesgue-Stieltjes integrals (so that the cumulative distributive function will determine the probability measure to be used). For most applications measure theory is not required, and the usual Riemann integral is sufficient.

There is 1 other function associated with every random variable X. If for every real number x we calculate the probability $F(x)$ that the random variable X is less than or equal to the number x, we get a function $F(x)$, which we call the *cumulative distribution function* of X and which is abbreviated *cdf*.

(The word "cumulative" is frequently omitted.) It is easily shown that $F(x)$ is a real, nonnegative, nondecreasing point function, continuous from the right, with $F(-\infty) = 0$ and $F(\infty) = 1$. The domain is the set of extended real numbers that include $(-\infty,\infty)$, and the range is the unit interval $[0,1]$. If $F(x)$ has a derivative (which is guaranteed if $F(x)$ is "absolutely continuous"), there exists another function, $f(x)$, called the *probability density function* (or *frequency distribution* or *probability law*) of X, abbreviated *pdf*. The pdf gives the "probability" at each value of the random variable. The pdf is a function of x with 4 properties: (1) the curve representing $f(x)$ lies entirely above the x-axis so that $f(x) \geq 0$, (2) the area under the entire curve is 1, i.e., $\int_{-\infty}^{\infty} f(t)dt = 1$, (3) the cdf is the area under the curve up to x, i.e., $F(x) = \int_{-\infty}^{x} f(t)dt$, and (4) $f(x)$ is the derivative of $F(x)$. Both the cdf and pdf occupy very important places in statistics because they enable us to calculate probabilities of events. For the normal distribution, the pdf is the familiar bell-shaped curve symmetric about the mean μ and having points of inflection 1 standard deviation in either direction from μ. The corresponding $F(x)$ is a "sigmoid" or S-shaped curve symmetric about the mean and bounded by $y = 1$ and $y = 0$.

If $F(x)$ is continuous but not absolutely continuous, X is said to have a *singular distribution*. If $F(x)$ is a step function (i.e., it assumes at most a countable number of values), the random variable X is said to be a *discrete random variable*. If $F(x)$ assumes a "continuum" of values, the random variable X is a *continuous random variable*. The adjective "continuous" used here does *not* imply that X is a continuous function in the mathematical sense since we do not define or even talk about the "limit" of the function $X(\omega)$ as ω approaches some value.

We may also define X to be a *discrete random variable* if its range is a finite or countable set and *continuous* if the range is a continuum. The values taken on by X when X is discrete are called *mass points*, and the density function is usually denoted by $p(x)$ rather than $f(x)$. In that case the function $p(x)$ is called the *probability mass function* and represents the probability that $X = x$. For all mass points x_i, $p(x_i) \geq 0$ and $\Sigma p(x_i) = 1$. For a continuous random variable, such as a man's height, the probability that X is equal to 72 inches (by which we mean 72 followed by more than 100 zeros) is obviously zero; hence we express the approximate probability that X lies in some *small* interval Δx, containing x, by $f(x)\Delta x$.

If all the probability is concentrated at 1 point, X is said to be a *degenerate random variable*. There are random variables that are a mixture of discrete and continuous random variables. It is also useful to define the "empirical" distribution function with reference to a *simple random sample* $X_1, \ldots, X_n$ in which the X's are independent and identically distributed random variables. The function $F_n(x) = i_x/n$, where i_x = the number of $X_i \leq x$, is called the *empirical* (or sample) *cumulative distribution function.* $F_n(x)$ is a step function that takes on values of 0, $1/n$, $2/n$, . . . , 1.

A very important concept in dealing with random variables is that of a *weighted average* of the values of the random variables. The "weight" associated with each value assumed by the random variable is the probability of being equal to that value, and the weighted average will be called an *expected value.* For a discrete random variable X, the *expected value* is $E(X) = \Sigma x_i p(x_i)$. For the continuous case the *expectation or expected value* is $E(X) = \int_{-\infty}^{\infty} x f(x) dx$. If the pdf $f(x)$ does not exist, $E(X) = \int x dF$. By convention we use the same limits $(-\infty, \infty)$ for all $f(x)$ by defining $f(x)$ to be zero except where it is positive. More generally, if $g(x)$ is a function of x, $E(g(X)) = \int_{-\infty}^{\infty} g(x) f(x) dx$ if X is continuous and $\Sigma g(x_i) p(x_i)$ if discrete. Expected values are used extensively.

If k is a positive integer, $E(X^k)$ is the *k-th raw moment* of X, $E(|X|^k)$ is the *k-th absolute moment* of X, $E[(X-c)^k]$ is the *k-th central moment about c* or simply the *k-th moment about c*, $E(X - E(X))^k$ is the *k-th central moment* of X, $E[|X - E(X)|^k]$ is the *k-th absolute central moment*, and $E[X(X-1), \ldots , (X-k-1)]$ is the *k-th* factorial moment of X. The first moment, $\mu = E(X)$, is the *mean* of X; the second moment about the mean, $\sigma_x^2 = E[X - E(X)]^2$ is the *variance* of X and its positive square root σ_x is the *standard deviation* of X. The quantity σ_x/μ_x is the *coefficient of variation*, or *relative standard deviation.* Its square is called the *relative variance* or the *rel-variance.* If Y is a second random variable, $E[(X - E(X)) (Y - E(Y))]$ is called the *covariance* of X and Y, denoted by $cov(X,Y)$ or σ_{xy}. If σ_x^2 and σ_y^2 are finite and positive, the *correlation coefficient* of X and Y is $\rho(X,Y) = cov(X,Y)/\sigma_x \sigma_y$, which lies between -1 and 1. X and Y are *uncorrelated* if $cov(X,Y) = 0$. Letting μ_k denote the *k-th* central moment, the *moment-ratios* $\beta_1 = \mu_3^2/\mu_2^3$ and $\beta_2 = \mu_4/\mu_2^2$ have been used respectively as measures of *skewness* (asymmetry) and *kurtosis.* It was formerly thought that kurtosis measured "peakedness," but it probably has more to do with tail behavior.

The extent to which the first 2 moments, μ and σ^2, characterize the distribution of the random variable is a result known as *Tchebycheff's Inequality* (for which there are at least a dozen spellings): The probability that the random variable X lies between $\mu - k\sigma$ and $\mu + k\sigma$ is greater than $1 - 1/k^2$.

Another valuable inequality concerns the union of events. If A is the finite union of events A_1, A_2, . . . A_N, the probability of A, $P(A)$, is the sum of the probabilities of the A_i taken 1 at a time minus those taken 2 at a time plus those taken 3 at a time, etc. The *Bonferroni inequalities* then follow: $P(A) \leq \Sigma P(A_i)$; $Pr(A) \geq \Sigma P(A_i) - \Sigma P(A_i A_j)$ (where in the last term $i < j$); $Pr(A) \leq \Sigma P(A_i) - \Sigma P(A_i A_j) + \Sigma P(A_i A_j A_k)$; etc.

The moments defined above are more properly called *population moments.* Given a random sample $X_1, \ldots , X_n$ of size n, each population moment can be estimated by the corresponding *sample moment*. In particular, the k-th *sample moment* is $\sum X_i^k/n$, so that the first sample moment or *sample mean* is $\overline{X} = \sum X_i/n$ while the *sample variance* is $\Sigma(X_i - \overline{X})^2/n$. The sample variance, in that form, is biased, and it is customary to use the unbiased sample variance $S_x^2 = \sum(X_i - \overline{X})^2/(n - 1)$ in most applications. The *sample coefficient of variation* or *sample relative standard deviation* is $S/\overline{X}$, the *sample covariance* is $S_{xy} = \sum(X_i - \overline{X})(Y_i - \overline{Y})/(n-1)$, and the *sample correlation coefficient* if $r = S_{xy}/S_x S_y$. Less used are the *geometric mean* (the n-th root of the product of the data points) and the *harmonic mean* (the reciprocal of the mean of the reciprocals of the data points). The log of the geometric mean is the mean of the logs of the data points, and by analogy the log of the *geometric standard deviation* has been defined as the standard deviation of the logs of the data. The geometric measures are not very useful in understanding the problem and not recommended. What should be done is to transform the data to logarithms and think only about means and standard deviations of the transformed data. The failure to observe the distinction between population moments and sample moments has led to numerous technical errors in the scientific literature.

Several functions "generate" moments. The function $f(t) = E(e^{tX})$, if it exists for an interval of t, is called the *moment generating function* of X, since its m-th derivative, evaluated at zero, is the m-th moment of X. The function $g(t) = E(e^{itX})$, where $i = \sqrt{-1}$, always exists and is called the *characteristic function* of X, since its m-th derivative, divided by i^m and evaluated at the origin is the m-th moment of X. $H(t) = E(t^X)$ is called the *factorial moment generating function*, since its k-th derivative, evaluated at 1, yields the k-th factorial moment. If X is discrete and $H(t)$ is written as a power series in t, the coefficient of t^X is $p(x)$; hence $H(t)$ is also called the *probability generating function*. The logarithm of the moment generating function is termed the *cumulant generating function* of X, and the coefficient of $t^k/k!$ in the Taylor Series expansion of the cumulant generating function is called the *k-th cumulant* or *semi-invariant* of X.

The *normal distribution* is the most important in statistics. The pdf is $f(x;\mu,\sigma) = [\sigma(2\pi)^{1/2}]^{-1}\exp(-(x - \mu)^2/2\sigma^2)$, where μ is the mean and σ^2 is the variance. The pdf thus represents a *family of distributions*; there is 1 member of the family for each value of μ and σ^2. The constants μ and σ^2 are called *parameters*, since they can be used to index the family. We use the notation $f(x;\theta)$ to represent a general pdf and let θ stand for 1 or several parameters while x represents the values taken on by 1 or more random variables. The parameters do not always represent the mean or the variance, as is the case with the normal. A list of the most useful distributions with their pdf's, means, and variances is given in Chapter 3. The pdf given above for the normal gives the *functional form* for all normals, but the parameters are unknown. When more than 1 random variable is involved, there is a joint pdf, joint cdf, marginal distributions, etc. They are treated in Chapter 13.

REFERENCES

The theory of random variables and probability is discussed to some extent in nearly every text in statistics. For good definitions I recommend Mood, A. M., Graybill, F. A., and Boes, D. C. 1974. *Introduction to the Theory of Statistics.* 3rd ed. New York: McGraw-Hill. For the theoretician, Ash, R. B. 1972. *Real Analysis and Probability.* New York: Academic Press should do.

3

Some Useful
Distributions

The *normal* or *Gaussian distribution* is the most widely used distribution in the field of statistics for 3 reasons: (1) numerous natural phenomena, or transformations of phenomena, are nearly normal, (2) sums of random variables from nonnormal distributions are near normal, and (3) the normal is the limiting distribution of many distributions (binomial, chi-square, Poisson, etc.).

A random variable X is said to be *normally* distributed with mean μ and variance σ^2 if the pdf is $f(x) = (2\pi\sigma^2)^{-1/2}\exp(-(x-\mu)^2/2\sigma^2)$. It is written $X \sim N(\mu,\sigma^2)$. If $Y \sim N(0,1)$, Y is said to have a *unit normal* or *standard normal* distribution; the cumulative distribution function (cdf) is denoted by $\Phi(y)$ and represents the area under the pdf from $-\infty$ to y. The normal has the familiar bell-shaped curve, symmetric about the mean, with points of inflection at $\mu \pm \sigma$. Any normal variate, $X \sim N(\mu,\sigma^2)$, can be transformed to a $N(0,1)$ by subtracting the mean and dividing by the standard deviation, i.e., $(X-\mu)/\sigma \sim N(0,1)$. Tables of the cdf for the standard normal are to be found in nearly every statistical textbook. Linear combinations of normals are normal: If $X \sim N(\mu_x,\sigma_x^2)$ and $Y \sim N(\mu_y,\sigma_y^2)$ and X and Y are independent, $aX + bY \sim N(a\mu_x + b\mu_y, a^2\sigma_x^2 + b^2\sigma_y^2)$.

Let $X_1, X_2, \ldots X_n$ be independent unit normals. Then ΣX_i^2 has, by definition, a *chi-square distribution* with n degrees of freedom, written $\Sigma X_i^2 \sim \chi^2(n)$. The pdf is $f(x) = (2^{n/2}\Gamma(n/2))^{-1} x^{n/2-1}\exp(-x/2)$, for $x > 0$ and $n = 1,2, \ldots$. The mean is n and the variance is $2n$. Also, if $\overline{X} = \Sigma X_i/n$ and $S^2 = \Sigma(X_i - \overline{X})^2/(n-1)$, $(n-1)S^2/\sigma^2 \sim \chi^2(n-1)$. Further, if $Y \sim N(0,1)$ and $Z \sim \chi^2(n)$, and Y is independent of Z, $T = Y/\sqrt{(Z/n)}$ has, by definition, a *Student's t distribution* or *t-distribution* with n degrees of freedom. That distribution, with parameter $n = 1,2, \ldots$ has pdf $f(x) = \dfrac{\Gamma[(n+1)/2](1+x^2/n)^{-(n+1)/2}}{\Gamma(n/2)\sqrt{(n\pi)}}$, written $T \sim t(n)$.

Also, if $U \sim \chi^2(n)$ and is independent of $V \sim \chi^2(m)$, then $Z = (U/n)/(V/m)$ has, by definition, an *F-distribution* with n and m degrees of freedom, sometimes called *Snedecor's F-* or the *variance-ratio* distribution. The pdf is $f(x) = \dfrac{\Gamma[(m+n)/2](n/m)^{n/2}x^{n/2-1}[1+(n/m)x]^{-(m+n)/2}}{\Gamma(m/2)\Gamma(n/2)}$. A variable Y has a *log-normal distribution* if $X = \ln Y$ has a $N(\mu,\sigma^2)$ distribution, and the pdf is $f(y) = [y\sigma(2\pi)^{1/2}]^{-1}\exp[-(\ln y - \mu)^2/(2\sigma^2)]$. The mean of Y is $\exp(\mu+\sigma^2/2)$, and the variance is $\exp(2\mu+2\sigma^2) - \exp(2\mu+\sigma^2)$. Particle sizes usually have lognormal distributions, and many mechanical and electric devices have lognormal lifetimes. If $X \sim N(\mu,\sigma^2)$, $Y=|X|$ has a *folded normal distribution*. In the special case of $\mu = 0$, Y has a *half-normal distribution*. The bivariate normal with zero correlation and equal standard deviations is the *circular normal*.

If $X_1, X_2, \ldots X_n$ are independent and $X_i \sim N(\mu_i,1)$, then ΣX_i^2 has a *non-central chi-square distribution* with n degrees of freedom and *noncentrality parameter* $\delta = (\Sigma\mu_i^2)^{1/2}$. If $X \sim N(\mu,1)$ and $U \sim \chi^2(n)$ and independent of X, then $X/(U/n)^{1/2}$ has a *noncentral t-distribution* with noncentrality parameter μ. If U_1 and U_2 are independent and U_1 is noncentral chi-square with noncentrality parameter δ and n_1 degrees of freedom, while U_2 is chi-square with n_2 degrees of freedom, $(U_1/n_1)/(U_2/n_2)$ has a *noncentral F-distribution* with noncentrality parameter δ.

Variables that are not normal can frequently be transformed to normality. The distribution of the sample correlation coefficient r is rather intractable, but for the *z-transformation*, $Z = (1/2)\ln[(1+r)/(1-r)]$ is approximately normal with mean $(1/2)\ln[(1+\rho)/(1-\rho)]$ and variance $1/(n-3)$, where ρ is the population correlation coefficient. In other cases, such as a binomial proportion X, the variance of the observations is not a constant, but a function of the mean; in that case the *arcsine transformation*, $Y = 2\sqrt{n} \arcsin \sqrt{X}$, has variance 1 and is thus called a *variance stabilizing transformation*. If the variance is proportional to the expected value with constant of proportionality k, (i.e., Var $X = kE(X)$, the *square root transformation*, $Y = \sqrt{X}$, gives a constant variance. If the standard deviation is proportional to the expected value, the *logarithmic transformation*, $Y = \ln X$, has a constant variance k^2. If the standard deviation is proportional to the square of the expected value, the *reciprocal transformation*, $Y = 1/X$, has constant variance k^2.

A Student's t variate with 1 degree of freedom or the ratio of 2-unit normals has a *Cauchy distribution* with parameters α and $\beta > 0$ with pdf $f(x) = [\pi\beta(1 + (x - \alpha)^2/\beta^2)]^{-1}$. Not one of the moments of that distribution is finite, but α is a location parameter and β is a scale parameter. The distribution is symmetric about α, which is the medium.

The *uniform* or *rectangular distribution* is defined on the interval [a,b] and has pdf $f(x) = 1/(b-a)$. The mean is $(b+a)/2$, and the variance is $(b-a)^2/12$. The probability of X being anywhere in the interval is the same. The average of 2 uniforms has a *triangular distribution*, so named because the pdf is shaped like a triangle. If X is any continuous variate with cdf $F(x)$, then the variable $Y = F(X)$ is uniform on the interval [0,1], a result known as the *probability integral transformation*.

The *gamma distribution* has parameters r and λ, both positive, and pdf $f(x) = [(\Gamma(r)]^{-1}\lambda^r x^{r-1}\exp(-\lambda x)$, with $x \geq 0$, mean $= r/\lambda$, and variance $= r/\lambda^2$. With $\lambda = 1$, the pdf is in *standard form*. The cdf of the standard gamma is $[\Gamma(r)]^{-1}\int_0^x t^{r-1}e^{-t}dt$, and the integral alone is referred to as the *incomplete gamma function*. If $r = 1$, we have an *exponential distribution* with pdf $f(x) = \lambda\exp(-\lambda x)$ for $x \geq 0$. The sum of independent exponentials has a gamma distribution. If the number of events happening has a Poisson distribution, the *waiting time* from time zero to the r-th event has a gamma distribution, and the waiting time to the first event has an exponential distribution. If r is a positive integer, the gamma is called an *Erlang distribution*. With $\lambda = 1/2$ and $r = n/2$, the gamma is a chi-square distribution with n degrees of freedom. If Y is uniform on $(0,1)$, $Z = -\ln Y$ has an exponential distribution and the sum of k such variables has a gamma distribution with $r = k$ and $\lambda = 1$. If X has a chi-square distribution with n degrees of freedom, then $Y = \sqrt{X}$ has a *chi-distribution* with pdf $f(x) = [2^{n/2-1}\Gamma(n/2)]^{-1} x^{n-1} \exp(-x^2/2)$. A chi-distribution with $n = 2$ is a *Rayleigh distribution*. The *exponential distribution* given above plays a central role in Reliability Theory and life-testing. When events occur at random in time, they have an exponential distribution. That means that the probability of failure of a device with such a distribution is the same, no matter how old the device is. If X is a variable such that $Y = (X - \theta)^c$ is exponential, X has a *Weibull distribution*. If $Y = e^{-X}$ has an exponential distribution, X has an *extreme-value distribution*. If a variable X has pdf $f(x) = (2\sigma)^{-1} \exp(-|x-\theta|/\sigma)$, X has a *double exponential* or *Laplace distribution*. A variable with cdf $F(x) = [1 + \exp(-(x-\alpha)/\beta)]^{-1}$ or pdf $f(x) = (4\beta)^{-1}\text{sech}^2[(x-\alpha)/2\beta]$ has a *logistic* or *sech-squared distribution*. The logistic is the limiting distribution of the standardized mid-range and is used in describing growth and is a substitute for the normal.

A *beta distribution* with parameters a and b, both positive, has pdf $f(x) = x^{a-1}(1-x^{b-1})/B(a,b)$, where x lies between 0 and 1 and $B(a,b)$ is the *Beta function*. If X_1^2 and X_2^2 are independent chi-square variables with n_1 and n_2 degrees of freedom, $V = X_1^2/(X_1^2 + X_2^2)$ is beta with $a = n_1/2$ and $b = n_2/2$. A beta with $a = b = 1/2$ is an *arcsine distribution*, and a beta with $a = b = 1$ is the *uniform distribution*. The beta distributions are defined on a finite interval and are zero elsewhere. The cdf of the standard beta is $[B(a,b)]^{-1} \int_0^x t^{a-1}(1-t)^{b-1}dt$, and the integral itself is called the *incomplete beta function*. The *Dirichlet distribution* is the multivariate extension of the beta distribution.

A variable X is said to have a *Weibull distribution* if the pdf is $f(x) = acx^{c-1}\exp(-ax^c)$. Numerous mechanical and electrical devices have lifetimes that have a Weibull distribution. The *Pareto distribution* has parameters r and A, both positive, with pdf $f(x) = rA^r/x^{r+1}$ for $x \geq A$. It pertains mostly to the distribution of incomes.

An *extreme-value distribution* has 3 types. The Type I has cdf $F(x) = \exp[-\exp(-(x-\xi)/\theta)]$; Type II has cdf $F(x) = \exp[-((x-\xi)/\theta)^{-k}]$, and Type III has cdf $F(x) = \exp[-((\xi-x)/\theta)^k]$. Types II and III can be transformed to a Type I, so it is necessary to consider only Type I, sometimes called the *log-Weibull distribution*. The Type III distribution with $(-X)$ is also a Weibull distribution. The limiting distribution of the smallest or largest observation in a random sample has an extreme-value distribution, and for that reason it is used in predicting floods, earthquakes, extreme aircraft loads, etc.

The *Inverse Gaussian* or *Wald distribution* in standard form has pdf $f(x|\mu,\lambda) = (\lambda/2\pi x^3)^{1/2}\exp(\lambda(x-\mu)^2/2\mu^2 x)$. It is so named because the cumulant generating function is the inverse of that of the normal distribution. If $1/X$ has a normal distribution, the distribution of X has also been called an *inverse normal* or *inverse Gaussian*.

A number of discrete distributions are of considerable interest. A *Bernoulli trial* is one that has only 2 outcomes, which we label as 1 and 0, the 1 occurring with probability p and the 0 with probability $1-p$. The toss of a coin is 1 example. The *Bernoulli distribution* has pdf equal to p if $x = 1$ and equal to $1 - p$ if $x = 0$. It gives rise to 3 discrete probability distributions. The *binomial distribution* has parameters n and p, where $n = 1, 2, \ldots, 0 \leq p \leq 1$, and $p(x) = \binom{n}{x} p^x(1-p)^{n-x}$, $x = 0, 1, \ldots, n$ represents the probability of x "successes" in n independent Bernoulli trials, and where the probability of a success is p. The mean is np, and the variance npq. The binomial distribution was so named because $p(x)$ is 1 term in the expansion of $(p + q)^n$, where $p + q = 1$. The *negative binomial distribution* gets its name from a similar expansion where the exponent is negative. The pdf is $p(x) = \binom{r+x-1}{x}p^r q^x$, the mean is rq/p, and the variance is rq/p^2. In that distribution X is the *waiting time* until the r-th "success." For the special case of $r = 1$, the waiting time until the first success, the negative binomial reduces to $p(x) = pq^x$ with mean q/p and variance q/p^2. That special case is called the *geometric distribution* or *Pascal distribution*, since the values that the pdf takes on are the terms of a geometric series. As n gets large or for $p \simeq 1/2$ or for $npq > 9$, the binomial distribution is near-normal and can be approximated by a $N(np, npq)$ distribution, called the *normal approximation to the binomial*. For small p and large n, the probabilities may also be approximated with a Poisson distribution with parameter $\lambda = np$. The binomial is used to construct *nonparametric tolerance limits*, which require either (1) the sample size n so that at least a proportion of the population is between $X_{(r)}$ and $X_{(n+1-m)}$ with probability $1 - \alpha$ or more or (2) for fixed n the values of r and m in (1). $X_{(r)}$ is the r-th order statistic of the X's. The *multinomial distribution* is a generalization of the binomial in which there are k mutually exclusive outcomes, $A_1, A_2, \ldots A_k$, with probabilities $p_1, p_2, \ldots p_k$, and $\Sigma p_i = 1$. In a sequence of n trials, $n = \Sigma x$, the probability of getting outcome A_1 exactly x_1 times, outcome A_2 exactly x_2 times, etc., is $p(x_1, \ldots x_k) = \dfrac{n!}{x_1! x_2! \ldots x_k!} p^{x_1} \ldots p^{x_k}$, which is the pdf of the multinomial.

The *Poisson distribution* with parameter $\lambda > 0$ has pdf $p(x) = e^{-\lambda} \lambda^x / x!$ for $x = 0, 1, 2, \ldots$ The mean and the variance are equal to λ. If, in the binomial, $n \to \infty$ and $p \to 0$ in such a way that np remains a constant, say λ, the binomial probabilities approach the Poisson probabilities with $\lambda = np$. The binomial probabilities are difficult to calculate; hence the Poisson probabilities are used as a substitute. As λ gets large, the Poisson approaches a normal and the normal may be used to approximate the Poisson probabilities. The chi-square distribution is used to place confidence limits on a Poisson parameter.

Any random phenomenon in which we count the number of events in a certain time interval of length t has a Poisson distribution with parameter $\lambda = vt$ if the probability of 1 event happening in a small interval $(t, t, + h)$ is approximately equal to vh (proportional to the length of the interval) and the probability of more than 1 event happening in the same interval is negligible in comparison with 1 event happening there and the number of occurrences in nonoverlapping time intervals is independent. The quantity v is the number of occurrences per unit time called the *mean occurrence rate*. The Poisson serves as a model for radioactive decay, the number of flaws per yard of material, the number of typographical errors per page, the number of telephone calls per hour, etc.

The *hypergeometric distribution* has parameters N, n, p with $N = 1, 2, \ldots$; $n = 1, 2, \ldots N$; $p = 0, 1/N, 2/N, \ldots 1$; and pdf $p(x) = \binom{Np}{x} \binom{N(1-p)}{n-x} / \binom{N}{n}$, with $x = 0, 1, \ldots n$. The mean is np, and the variance is $(npq)(N-n)/(N-1)$. If we draw a sample of size n, without replacement, from a population of N

elements of which Np are defective, the probability that the sample contains exactly x defectives is $p(x)$. That distribution is used extensively in quality control to determine the probability of accepting a lot of size N based on a sample of size n from the lot.

Two *systems of distributions* have been devised that contain a wide variety of continuous distributions and can be used to approximate continuous distributions of almost any type. The first is the *Pearson system*, which is divided into 7 types and includes the normal, gamma, chi-square, beta, uniform, and t-distributions. The members of that system have a pdf that satisfies the differential equation $f'(x)/f(x) = -(a + x)/(c_0 + c_1 x + c_2 x^2)$. For any distribution with known moments, it is possible to obtain a useful approximation of the pdf in terms of the unit normal density, the moments, and the Hermite polynomials (a convenient way to express the derivatives of the normal). That expansion is called the *Gram Charlier Expansion*. Another expansion of similar type in terms of the derivatives and cumulants is the *Edgeworth Expansion*. A form of the latter, called the *Cornish-Fisher Expansion*, is used to tabulate significance points of certain distributions by making the first few moments of the distribution match those in the expansion.

A second system of distributions is the *Johnson System*. Johnson gave 3 transformations that lead to a unit normal. X has a *Johnson's S_L distribution* if $Z = \gamma + \delta \log X$ has a unit normal distribution. X has a *Johnson's S_B distribution* if $Z = \gamma + \delta \log[X/(1 - X)]$ has a unit normal distribution, and X has a *Johnson's S_U distribution* if $Z = \gamma + \delta \sinh^{-1} X$ has a unit normal distribution. The subscripts $L, U,$ and B stand for logarithmic, unbounded, and bounded distributions. There is exactly 1 member of the Johnson system for every possible combination of skewness and kurtosis.

A *contagious distribution* is a *mixture of distributions* whose density is obtained by taking a weighted average (the weights add to 1) of two or more distributions. A *compound distribution* is constructed by allowing some of the parameters of the distribution to have a probability distribution. A mixture may thus be thought of as a compound distribution. There are many compound distributions with applications. A *Polya* or *Polya-Eggenberger distribution* is a binomial in which the parameter p has a beta distribution. If there are n white balls and m black balls in an urn and 1 ball is drawn and replaced by s balls of the same color and if that procedure is repeated N times and X is the total number of times a white ball is drawn, X has a Polya distribution. If a variable X has a binomial distribution with parameters N and p and if N/n has a Poisson distribution, X has a *Poisson-binomial* distribution. If X has a negative binomial distribution with parameters N and p and if N/k has a Poisson distribution, X has a Poisson-Pascal distribution. A *modified distribution* is one in which there are more zeros than would be expected from a Poisson distribution. The result is a *Poisson with added zeros*. The term "modified" could also be used to describe other types of modifications. If a distribution is *truncated* (i.e., its values below a and above b are not known), the area A under the pdf from a to b does not add to 1 but the truncated portion may be made into a valid *truncated distribution* by dividing the cdf by A. If the argument t of a probability generating function $g_1(t)$ of a discrete distribution is replaced by the probability generating function $g_2(t)$ of another distribution, then $g_1(g_2(t))$ is also a generating function and the distribution whose probabilities it produces is called a *generalized distribution*.

Given a random sample of size n from an unknown distribution, the null hypothesis that the unknown distribution is a specified distribution is tested with a *goodness-of-fit test*. The *Kolmogorov-Smirnov test* uses as a statistic the largest absolute difference between the empirical cdf $S(x)$ and the theoretical or specified cdf $F(x)$. The parameters of the specified distribution must be given; they cannot be estimated from the data. One cannot use that test to decide whether a set of observations comes from a "normal" distribution, but one can decide whether they come from a $N(3,5)$ distribution. If the statistic exceeds the tabulated critical value $d_{n,\alpha}$, we decide that the observations are not from $F(x)$. In some cases the parameters can be estimated from the data provided that one uses a different set of critical values; in the normal case the test is then known as a *Lilliefors test*. By placing a band of width $d_{n,\alpha}$ above and below $S(x)$, we have a confidence band for the distribution function $F(x)$.

The *Cramer-Von Mises test* is much like the Kolmogorov-Smirnov test with this difference: The average of i/n and $(i-1)/n$ (the values of $S(x)$ just after and just before the jump at the i-th point) are used instead of $S(x)$ to obtain the differences. The statistic used is then $1/(12n)$ plus the sum of squares of the n differences. The *Anderson-Darling test* uses as a statistic the integral $\int_{-\infty}^{\infty}\{[S(x)-F(x)]^2[F(x)(1-F(x))^{-1}dF$, but it is not widely used.

In the popular *chi-square goodness-of-fit test*, the parameters of $F(x)$ are also required to be specified. We divide up the abscissa of the pdf $f(x)$ and calculate the probability p_i for each of the intervals. Then np_i is the number of observations expected to fall in the i-th interval and n_i the observed number. The test statistic $\chi^2 = \sum\dfrac{(np_i - n_i)^2}{np_i}$ is then used to compare observed and expected frequencies in each category. Though widely utilized, the test suffers from a dependence on a subjective decision of how many intervals to use. In practice estimated parameters are often employed with effects that are not overly serious.

There are a number of tests of normality, some based on the above, the best of which is the *W-test* (named for Wilk and Shapiro). The statistic is the ratio of the squared linear estimator of the standard deviation to the sum of squares of deviations from the sample mean (a quadratic estimator). An older test of normality used as a statistic *Geary's ratio*, which was the mean deviation divided by the standard deviation. Finally, there are two-sample versions of the Kolmogorov-Smirnov test used to decide whether 2 unknown distributions are identical. The *Smirnov test* uses as a statistic the largest absolute difference between the 2 empirical cdf's. The *Cramer-Von Mises* two-sample test uses the sum of squares of the differences between the two empirical cdf's multiplied by the factor $mn/(m+n)^2$.

A random variable X is *infinitely divisible* if X can be written as a sum of n i.i.d. (independent and identically distributed) random variables for every positive integer n. X is then said to have an *infinitely divisible distribution*. The normal, gamma, Poisson, and Cauchy random variables are infinitely divisible.

We need to speak here about the subject of convergence in order to define the Central Limit Theorem and related terms. If we have a sequence of *points*, $X_1, X_2, \ldots$, we say that the sequence *converges* to a point p if there is an N such that for all $n > N$, X_n is closer than ϵ to p, where ϵ is some small positive number. A sequence of *functions* $f_1(x), f_2(x), \ldots$ *converges pointwise* to a function $f(x)$ if for every given x_0 the sequence of points $f_1(x_0), f_2(x_0), \ldots$ converges. That means that for any given x and ϵ, there is an N such that for all $n > N$, $|f_n(x) - f(x)| < \epsilon$. The value of N depends on the value of x and ϵ chosen. If 1 value of N will work for the given ϵ and for *all* x, the sequence *converges uniformly* to $f(x)$. We say that a sequence coverges *almost everywhere* to $f(x)$ if it converges for every x except on a set of measure zero. (We regret not being able to state that more simply—the values of x on which it fails to converge are insignificant.) Thus uniform convergence implies convergence almost everywhere. *Convergence in probability* or *measure* or *stochastic convergence* is still weaker and is implied by convergence almost everywhere. A sequence of functions $f_1(x), f_2(x), \ldots$ converges in probability to $f(x)$ if for each ϵ the probability that $|f_n(x) - f(x)] \geq \epsilon$ tends to zero as $n \to \infty$ (or the probability that $|f_n(x) - f(x)| < \epsilon$ tends to unity as $n \to \infty$).

The *weak law of large numbers* is as follows: Let $X_1, X_2, \ldots$ be independent but not necessarily identically distributed random variables, each with finite mean and variance. Let all the variances be less than some finite number M and let $S_n = X_1 + X_2 + \ldots + X_n$. Then $[S_n - E(S_n)]/n$ converges in probability to zero.

The *strong law of large numbers* states that if $X_1, X_2, \ldots$ are independent and identically distributed random variables with finite mean μ and $S_n = X_1 + X_2 + \ldots X_n$, then S_n/n converges to μ almost everywhere (except on a set whose probability is zero). There are a number of versions of that law, but all are statements about convergence almost everywhere. Included among the versions are the Bernoulli, Borel, Chebyshev, Khintchin, Kolmogorov, and Poisson laws of large numbers.

The Poisson distribution is occasionally called the *law of small numbers* simply because the random variable can assume many different values but with small probabilities.

The strong law of large numbers given above says nothing about the distribution of S_n. If the random variables in the sample are independent and identically distributed with finite mean μ and finite variance σ^2, the *Central Limit Theorem* states that $(S_n - n\mu)/\sigma\sqrt{n}$ is approximately normal with zero mean and unit variance. There are also many versions of that theorem. The fact that sums of random variables from any distribution tend to normality is what makes the normal distribution so important in statistics.

Monte Carlo simulation methods were invented at Los Alamos for coping with the difficult calculations associated with nuclear reactions. The name appropriately conveys an element of randomness and an element of pretense or mock-up. In mathematics the method has been used for difficult integration problems. If we have a known function, $f(x)$, whose graph lies entirely within a rectangle of area A and if we generate points at *random* within the rectangle (the points are equally likely to fall anywhere within the rectangle), we can calculate the fraction P of points that fall under the curve $f(x)$. The area under the curve is then A times P.

Nearly any system can be simulated if the components are well understood. If we were comparing the median, midrange, and arithmetic mean as estimators of the mean μ of a normal population, say, we would generate 1,000 samples, each of size n, from a normal distribution and calculate from each the median, mean, and midrange. In generating the samples, we *know* the mean μ and standard deviation σ of the population. We arrange the 1,000 sample values of each statistic in order, and that gives us a good estimate of the distribution of each statistic. We can also calculate the bias, standard deviation, and mean square error of each statistic and use them as a means of comparing the 3 estimators. The distribution and comparisons will depend on the sample size n; hence we will want to do the same thing for several sample sizes. That method is widely used, so we introduce some of the terminology associated with generating random variables from various distributions.

The generation of uniform random numbers (uniformly distributed on $(0,1)$) forms the basis of generating random variates from any other distribution. Most random number generators use a recursion formula to get the next random number X_{n+1} from the present one X_n. The formula is $X_{n+1} = aX_n + b \pmod{m}$ for $n \geq 0$; a, b, and m are integers chosen with efficiency in mind. The formula starts from a *seed*, x_0, which is given. (A number x, mod m, is the remainder after x is divided by m.) That formula is deterministic. There is nothing random in it; hence the generated numbers are called *pseudo-random numbers*. Given an X_i, we get the uniformly distributed numbers by taking $U_i = X_i/m$. The set of numbers generated eventually repeats itself; it has a cycle. The integer constants are chosen to make that cycle very large indeed. The generator is described as a *congruential generator* because the theory of congruences can be used to investigate its behavior. If b is zero, it is further called a *multiplicative congruential generator*; and if b is not zero, it is a *mixed congruential generator*.

For generating nonuniform random variates, let U_1 and U_2 be uniform on $(0,1)$ and independent. The *Box-Muller method* uses U_1 and U_2 to generate 2 independent standard normals: $N_1 = (-2 \ln U_1)^{1/2} \sin(2\pi U_2)$ and $N_2 = (-2 \ln U_1)^{1/2} \cos(2\pi U_2)$. In the *Polar-Marsaglia method*, we calculate $V_1 = 2U_1 - 1$ and $V_2 = 2U_2 - 1$ so that V_1 and V_2 are independent and uniform on $(-1,1)$. If $W = V_1^2 + V_2^2 \leqslant 1$, then $N_1 = V_1((-2log\ W)/W)^{1/2}$ and $N_2 = V_2((-2log\ W)/W)^{1/2}$ are independent standard normals.

For generating chi-square variates, let $U_1, U_2, \ldots U_m$ be uniform on $(0,1)$. Then $X = -2 \ln(U_1 U_2 \ldots U_m)$ is chi-square with $2m$ degrees of freedom and $X + Y^2$, where Y is a standard normal, is a chi-square with $2m+1$ degrees of freedom. For large degrees of freedom, the *Wilson-Hilferty transformation* $X = n[Y\sqrt{(2/9n)} + 2/9n + 1]^3$ is chi-square with n degrees of freedom, where Y is a standard normal.

Again, if U is $U(0,1)$ then $X = -(\ln U)/\lambda$ has an exponential distribution with parameter λ. Also, $Y = \alpha - \beta \ln[(1 - U)/U]$ has a logistic distribution with parameters α and β.

If $U_i \sim U(0,1)$, then $X = -(1/\lambda) \ln(U_1 U_2 \ldots U_k)$ has a gamma distribution with parameters k and λ. If X_1 is gamma with parameters 1 and p and X_2 is gamma with parameters 1 and q (where p and q are integers) and X_1 and X_2 are independent, then $X_1/(X_1 + X_2)$ has a beta distribution with parameters p and q.

To simulate binomial variates with parameters n and p, we generate n $U(0,1)$ variates, $U_1, U_2, \ldots U_n$. We let B_i be 1 if $U_i \leq p$ and zero otherwise. The sum of the B_i has the desired distribution. To generate Poisson variates, we generate E_i where E_i is exponentially distributed with parameter λ. Let S_k equal $E_i + E_2 + \ldots + E_k$. Let k increase until $S_k \leqslant 1$ and $S_{k+1} > 1$. The value of k then has a Poisson distribution with parameter λ.

In general, if we know the cdf $F(x)$, we can generate variates from it by using $X = F^{-1}(U)$, where $U \sim U(0,1)$. We do that by setting $U = F(x)$ and solving for x. That is the *inversion method*. For discrete random variables, we can frequently use the *table look-up method*. If X takes the values 0, 1, 2, . . . with probabilities $p_1, p_2 \ldots$, we form a table of integers 1,2, . . . N so that numbers 1 to k_1 correspond to $X = 0$, those from $k_1 + 1$ to k_2 to $X = 1$, etc. where $k_1/N = p_1$, etc. We then generate a discrete uniform variate on $(1,N)$ and from it obtain X_i.

For continuous random variables, the *rejection method* is frequently more efficient than the direct methods given above. If X has pdf $f(x)$ and if we could uniformly and randomly sprinkle points under $f(x)$, the points would be variates from the desired distribution. In practice we do the following: Find a pdf $h(x)$ with the same range as $f(x)$ and from which it is easy to simulate. Let $g(x)$ be some multiple of $h(x)$ such that $g(x)$ is just large enough to envelop $f(x)$. Simulate X from $h(x)$ and accept those X's from which another $U(0,1)$ variable is less than or equal to $f(x)/g(x)$.

The *composition method* is still another way of generating variates. If we wish to generate from $f(x)$ and if $f_1(x)$ is a pdf with similar shape from which it is easier to simulate, we can write $f(x)$ as a mixture: $f(x) = \alpha f_1(x) + (1 - \alpha)[(f(x - \alpha f_1(x))/(1-\alpha)]$, where $0 \leq d \leq 1$. We can then simulate from $f(x)$ by simulating from $f_1(x)$ 100α percent of the time and from $f_2(x)$ the expression in brackets, the rest of the time. We would like α as large as possible with the constraint that $f(x) - \alpha f_1(x) \geq 0$ so that $f_2(x)$ is a pdf. The 3 methods are often combined in simulating random variables.

References

A handy, readable book is Hastings, N. A. J., and Peacock, J. B. 1974. *Statistical Distributions*. Most statistical texts discuss the commonly used distributions. The authoritative reference in this area is the 4-volume set by N. L. Johnson and S. Kotz, published by Wiley, New York, N.Y. All of them have the title *Distributions in Statistics* with the following subtitles: *Discrete Distributions (1969), Continuous Univariate Distributions* (1970) (2 vols.), and *Continuous Multivariate Distributions* (1972).

4

Estimation and Hypothesis Testing

Estimation

If a layman were given the task of describing the average height of the population of white male college students at a certain university, he would no doubt take a sample from the population, measure the heights, and average the results. A quicker estimate would be the average of the smallest and largest heights in the sample. Another estimate would be the sample median (a number smaller than half the measurements and larger than the other half of them). Which of those estimates is "best" and why? If he were to take a different sample, the estimates would be different and he would need to express that *sampling* uncertainty in some way. This chapter defines the desirable properties of *point estimates* (which consist of 1 number). The uncertainty of the estimate resulting from sampling will be expressed in the form of *interval estimates* (which will consist of 2 numbers) of the quantity being estimated (in this case the height of men).

Suppose now that someone had claimed that the average height of black college males is greater than the average height of white college males. How could that conjecture about the parameters (mean heights) of the 2 distributions be tested? What is usually done is to get the average height of a sample of black males and the average height of a sample of white males. The difference in average heights is then compared with the variability within the 2 samples. If the difference in average height is greater than we would expect, we would decide that one group was taller than the other. That procedure is one instance of *hypothesis-testing*. Many other instances will be cited in this chapter.

There are 2 natural branches of statistical inference: *hypothesis testing* and *estimation*. We shall postpone hypothesis testing for now and concentrate on *estimation* of the parameters of some distribution. That type of estimation is called *parametric estimation*; it assumes that the functional form of the pdf is known, but the parameters are not. *Nonparametric estimation* refers to estimation of certain moments or quantiles that do not depend on, and do not require a knowledge of or an assumption of, the functional form of the pdf.

Estimation is further divided into *point* and *interval estimation*. In point estimation a random sample $X_1, \ldots, X_n$ is selected from a population whose pdf, $f(x;\theta)$, is assumed known (except for θ). The sample $X_1, X_2, \ldots, X_n$ consists of symbols for random variables. We let $x_1, x_2, \ldots, x_n$ be their actual or *realized values*. We devise some function, $T(X_1, \ldots, X_n)$, which is a formula that we think would estimate θ credibly. If θ were the mean of the normal distribution, say, θ could be estimated sensibly by the average of all the sample values, by the midrange of the values, or by the median. $T(X_1, \ldots, X_n)$ is a random variable called a *point estimator* of θ, and its realization, $t(x_1, \ldots, x_n)$ is a number called the *point estimate* of θ, a distinction worth preserving.

A *statistic* is any function of the sample values that does not contain any parameters. The arithmetic mean $\overline{X}$ is a statistic used to estimate the mean μ of the normal distribution, but $\overline{X} - \mu$ is not a statistic because μ is a parameter. A statistic is a random variable (as is every function of a random variable). We now wish to examine some desirable properties of a statistic. Suppose first that we know the functional form of the pdf, $f(x;\theta)$, but do not know the parameter θ. We obtain a sample $X_1, X_2, \ldots, X_n$ and from it calculate a statistic, S. The statistic summarizes or condenses the information found in the sample. If the statistic used contains just as much information about the parameter being estimated as the sample does, it is called a *sufficient statistic*. More precisely, we say that a statistic is a *sufficient statistic* for θ if and only if the conditional distribution of the sample given S does not depend on θ for any value s of S. If the distribution does not depend on θ, the sample contains no additional information about θ. A statistic can be tested for sufficiency in various ways. For some problems no single sufficient exists. There is always a set of statistics that is jointly sufficient. The set of statistics $S_1, S_2, \ldots, S_k$ is said to be *jointly sufficient* if and only if the conditional distribution of $X_1, X_2, \ldots, X_n$ given $S_1 = s_1, \ldots, S_k = s_k$ does not depend on θ. A set of jointly sufficient statistics is *minimal sufficient* if it is a function of every other set of sufficient statistics.

A second desirable criterion is to want an estimator T of θ that is *close*, in some sense, to θ. One measure of closeness is the *bias* of the estimator T, defined as $\theta - E(T)$. If $E(T) = \theta$, T is an *unbiased estimator*. Unbiasedness is only an average property, however, and a marksman aiming at a bull's eye could put every mark in the outermost circle and still, on the average, be unbiased. However, a marksman who can place every shot within a 1-inch circle but still miss the bull's eye substantially is said to be precise because his marks are repeatable. The desirable thing, of course, is to be both precise and unbiased. *Accuracy* is a general term for "closeness" to the "truth." However, a problem with that definition is that the "truth" is usually unknown and unknowable. We then have to compare our measurement with a "better" measurement, which will be called a "reference value," an "accepted value," a "standard value," or a "target value." Thus accuracy is not relative to the truth but to the better measurement. The *sample bias* is the accepted value minus the average value and is frequently used as a measure of accuracy. *Precision* is a general term that measures the scatter in the data using estimators, such as the variance, standard deviation, or range. A single measure including both precision and accuracy is *mean-squared error*, $E(T - \theta)^2 =$ variance + bias^2. Many people insist that "accuracy" should include good precision as well as small bias. Mean squared error is widely used as a standard for the goodness of an estimator, but it depends, in general, upon θ. Hence, estimators with uniformly (for all θ) minimum mean square error seldom exist. A *minimum variance* or *best estimator* in a class of estimators is one that has a smaller variance than any other estimator in the class. A *linear* estimator is a linear combination of the observations. A *BLUE* estimator is best in the class of linear unbiased estimators, and a *MULE* estimator is the same thing: a minimum variance unbiased linear estimator. Uniformly minimum mean square estimators are equivalent to uniformly minimum variance estimators in the class of unbiased estimators (the uniformity refers to all θ), and such an estimator is called a *uniformly minimum variance unbiased estimator* (UMVUE). The search is aided by the fact that under certain regularity conditions the variance of an unbiased estimator has a lower bound

$$1/nE\left[\frac{\partial}{\partial\theta} \ln f(x;\theta)\right]^2$$ called the *Cramer-Rao lower bound*. An unbiased esti-

mator that attains that lower bound is an *efficient* estimator, although that adjective is sometimes used to describe a function of a sufficient statistic. The ratio of the Cramer-Rao lower bound to the actual variance of an unbiased estimator is called the *efficiency* of the estimator. In general, the bound is not

attainable. The *Rao-Blackwell Theorem* states that, given an unbiased esti-
mator T and a sufficient statistic S, the estimator $T' = E(T|S = s)$ is unbiased
and has a smaller variance than T. A statistic T is *complete* if and only if the
only unbiased estimator of zero that is a function of T is the statistic that is
identically zero with probability 1. The *Lehmann-Scheffé Theorem* states that
an unbiased estimator that is a function of a complete sufficient statistic is a
unique UMVUE.

We now consider large sample properties of estimators. One desirable
property of an estimator of θ is that it gets closer to θ as the sample size
increases. Technically we say that a sequence $\{\hat{\theta}_n\}$ of estimators of θ, calculated in the same way except for the sample size n, is a *consistent estimator*
of θ if there exists an N such that for $n > N$ the probability that $|\hat{\theta}_n - \theta| <$
ϵ is greater than $1 - \eta$ for all positive ϵ and η, no matter how small. The
same sequence is an asymptotically *efficient estimator* of θ if (a) the distribution of $\sqrt{n}\,(\hat{\theta}_n - \theta)$ is asymptotically normal with mean zero and variance
σ^2 and (b) the variance of the estimator is less than that of other estimators
that satisfy condition (a). If one estimator has minimum variance V_1 and a
second estimator has variance V_2, the *efficiency* of the second estimator is $V_1/$
V_2. Another measure of relative efficiency, called *Bahadur efficiency*, is more
involved than we have space for discussing here.

A sequence of estimators is defined to be *mean squared error consistent*
if the limit of $E(\hat{\theta}_n - \theta)^2$ is zero for all θ as $n \to \infty$. That implies that both
the bias and the variance of θ approach zero as n gets large. Mean squared
error consistency implies simple consistency. An estimator that is *best asymptotically normal (BAN)* is one that is consistent and efficient.

One further desirable property of estimators is that of *invariance*. The
estimator θ is an *invariant estimator* of $\hat{\theta}$ for a certain class of transformations g if the estimator of $g(\theta)$ is $g(\hat{\theta})$. Maximum likelihood estimators owe
their importance to the fact that they are invariant for transformations with
single-valued inverses. For example, if $\hat{\sigma}^2$ is the maximum likelihood estimator of σ^2, then $\hat{\sigma}$ is the maximum likelihood estimator of σ. That is not a
general property of estimators: if $\tilde{\sigma}^2$ is an unbiased estimator of σ^2, $\tilde{\sigma}$ is not
an unbiased estimator of σ. In a related context, an estimator $\hat{\theta}$ is *location
invariant* if for every constant c and for all x_i, $\hat{\theta}(x_1 + c, \ldots, x_n + c) =$
$\hat{\theta}(x_1, \ldots, x_n) + c$ for all x_i and c; also, $\hat{\theta}$ is   *scale   invariant*   if
$\hat{\theta}(cx_1, \ldots, cx_n) = c(\hat{\theta}\,x_1 \ldots, x_n)$ for all x_i and positive c. For a random
variable X with density $f(x;\theta)$, θ is a *location parameter* if the distribution of
$X - \theta$ does not depend on θ, i.e., $f(x;\theta)$ can be written as $h(x - \theta)$ for some
$h(x)$. Specifically, θ will equal zero. For $\theta > 0$, θ is a *scale parameter* if the

distribution of X/θ is independent of θ. That requires that $f(x;\theta)$ be expressible as $(1/\theta)h(x/\theta)$ for some density $h(x)$ in which θ will equal 1. A *shape parameter* is a parameter in a pdf that is neither a location nor a scale parameter; it will affect the shape of the members of the family of distributions. The mean μ of a normal distribution is a location parameter, since $f(x;0,1) = (2\pi)^{-1/2}$ $\exp(-x^2/2)$ is a density. For the same reason, σ^2 is a scale parameter. In the gamma distribution the parameter r is a shape parameter.

Less frequently used terms for comparing closeness are the following: given estimators T and T' of θ, T' is *more concentrated* than T if $P(\theta - \lambda < T' \leqslant \theta + \lambda) \geqslant P(\theta - \lambda < T \leqslant \theta + \lambda)$ for all $\lambda > 0$ and all θ. An estimator is *most concentrated* if it is more concentrated than any other estimator. Most concentrated estimators do not generally exist. The estimator T' is *Pitman closer* than T if $P(|T' - \theta| < |T - \theta|) \geqslant 1/2$ for all θ. An estimator is *Pitman closest* if it is Pitman closer than any other estimator of θ.

We now classify several types of estimators according to the method used in finding them.

(1) The *method of moments estimator* is the estimator found by equating the population moments (which are functions of the parameters) to the sample moments and solving the resulting equations for the parameters.

(2) The *maximum likelihood estimator* is the estimator found by maximizing the *likelihood function* (the joint density of the sample) or the log likelihood with respect to the parameters. Maximum likelihood estimators are very important because they are invariant for transformations with single-valued inverses and under certain regularity conditions on the pdf are consistent, asymptotically normal, and achieve the Cramer-Rao lower bound on the variance.

(3) The *least squares estimator* is the estimator obtained by minimizing the sum of squares of deviations of the sample values from some function (of the parameters) that has been hypothesized as a fit for the data.

(4) The *minimum chi-square estimator* is the estimator found by minimizing the function $X^2 = \Sigma(n_j - np_j)^2/np_j$ with respect to the parameters. (It is assumed that the sample range has been partitioned into k intervals, that n_j is the number of observations falling into the j-th interval, and that np_j is the expected number of observations in the j-th interval. The probability p_j that an observation falls into the j-th interval is a function of the parameters of the assumed model with $\Sigma p_j = 1$ and $\Sigma n_j = n$.) If n_j is used in the denominator in place of np_j, the result is a *modified minimum chi-square estimator*.

(5) The *minimum distance estimator* is found by minimizing the distance between a member of an assumed class of cumulative distribution functions and the empirical cumulative distribution.

(6) *Admissible, Minimax, and Bayes Estimators*—It is sometimes assumed

that the parameter θ is behaving like a random variable with probability density function $g(\theta)$, called the *prior density of* θ. The conditional density of θ given the sample is called the *posterior density* of θ, denoted by $f(\theta|x)$. The *posterior Bayes estimator* of θ, with respect to the prior $g(\theta)$, is the expected value of θ given the sample.

Let the *loss function* $l(\hat{\theta},\theta)$ represent the "loss" incurred when $\hat{\theta}$ is used in place of θ. We require that $l(\hat{\theta},\theta)$ be real-valued, nonnegative, and zero when $\hat{\theta} = \theta$. Common loss functions are the *squared error loss*, $(\hat{\theta}-\theta)^2$, and the *absolute error loss*, $|\hat{\theta}-\theta|$. The *risk function* is the *average loss*, i.e., the expected value of the loss function. The risk function of squared error loss is the mean squared error and that of the absolute error loss is the mean absolute deviation. Given estimators T and T', we say that T is a *better* estimator than T' if the risk function for T is never greater than that for T' and is less than that for T' for at least 1 value of θ. An estimator T is *admissible* if and only if there is no other estimate better than T. An estimator is *minimax* if its maximum risk over all θ is less than or equal to the maximum risk of any other estimator. The *Bayes risk* of an estimator is the average (expected value) of the risk, the averaging being taken over the parameter space with respect to the prior distribution of θ. For a given loss function and prior density, the *Bayes estimator* of θ is the estimator with smallest Bayes risk. Bayesian methods provide a formal way of combining some notions about the uncertainties in the parameters (through the prior) with the data to obtain better information about the parameters (expressed through the posteriors). The prior $g(\theta)$ is not intended, however, to have a frequency interpretation. If we postulate a prior $h(\theta)$ that *does* have a frequency interpretation and that can be estimated (or partly estimated) from previous data, use of $h(\theta)$ in a Bayesian analysis constitutes an *Empirical Bayesian* analysis. Both Bayesian approaches are currently receiving a good deal of attention.

7. *Pitman estimators*—The *Pitman estimator for location* is the one that has the uniformly smallest mean squared error within the class of location invariant estimators. The *Pitman estimator for scale* is the one with uniformly smallest risk using the loss function $(\hat{\theta}-\theta)^2/\theta^2$. Those estimators have little practical importance.

Many estimators can be classified as 1 of 3 basic types: *An M-estimator* (for maximum likelihood) for a location parameter λ is a solution to the equation $\Sigma\psi(X_i - \lambda) = 0$ where X_i is the i-th data point and $\psi(x)$ is a defining equation. The asymptotically most efficient M-estimator uses $\psi(x) = -f'(x)/f(x)$, where $f(x)$ is the pdf and $f'(x)$ is the derivative of $f(x)$. *Andrews SINE estimator* uses $\psi(x) = \sin(x/2.1)$ if $|x| < 2.1\pi$ and $\psi(x) = 0$ otherwise. The latter estimator has been widely used in *outlier accommodation* (as opposed to *outlier detection*).

An *L-estimator* (for linear combinations) is a weighted average of the order statistics of the sample. For large samples the *i*-th weight is $h\left(\dfrac{i}{n+1}\right)\Big/\Sigma h\left(\dfrac{i}{n+1}\right)$, where $h(u) = g(F^{-1}(u))$, $g = -(d/dx)(f'(x)/f(x))$, and F is the cdf. In small samples the optimal weights are derived from the expected values and covariances of the order statistics.

An *R-estimator* (for ranks) is a solution of $\Sigma \, \mathrm{sgn}(X_i - \lambda) \, J^+[R(|X_i - \lambda|)/(n+1)] = 0$, where $J^+(u) = J(1/2 + u/2)$, $R(u)$ is the rank of u, and sgn is the signum function. The most efficient score function is $J(u) = (-f'(x)/f(x))(F^{-1}(u))$.

We now proceed to *interval estimation*, which consists of obtaining a pair of estimators to serve as the endpoints of a random interval in which the parameter will lie with some stated probability. An *interval estimator* for a parameter consists of 2 statistics or random variables: an *upper limit* (U) and a *lower limit* (L). With each interval there is an associated probability that the interval contains the parameter. If the probability statement involves only L or only U, we speak of a *one-sided interval*, while if it involves both L and U, it is called a *two-sided interval*. The probabilistic sense in which the intervals contain the unknown quantity is a hypothetical situation in which a large number of random samples are drawn and an interval constructed (in a prescribed manner) from each sample. A certain fraction of the intervals thus constructed would contain the unknown quantity. In practice, however, only 1 sample is used and 1 interval $(1,u)$ is constructed, where 1 and u are the realizations of L and U. The probability thus reflects the confidence we have in the procedure: Most of the time it yields an interval that will capture the parameter.

A $(1-\alpha)100$ percent *confidence interval* is an interval for an unknown parameter θ constructed from a sample in such a way that if the same method were used to construct a "large" number of such intervals from independent samples, $(1-\alpha)100$ percent of the intervals would contain the parameter θ. The term $(1-\alpha)$ is called the *confidence coefficient* or *confidence level*. A $(1-\alpha)100$ percent confidence interval, given by $L \leq \theta \leq U$, is equivalent to being $(1-\alpha)100$ percent "confident" that the true parameter is between L and U. That statement is called a *confidence statement*. It should be noted that no probability or confidence is attached to the estimate $(1,u)$. For any one given interval or realization, $(1,u)$, the probability that the parameter is between 1 and u is either zero or one. Confidence intervals are very important and are universally used as interval estimators. For a sample of size n, the $100(1 - \alpha)$ percent confidence interval on the mean μ of the normal distribution is $(\overline{X} \pm t_{n-1,1-\alpha/2} \, S/\sqrt{n})$ while the confidence interval for the variance σ^2 is $((n-1)S^2/\chi^2_{n-1,1-\alpha/2}, (n-1)S^2/\chi^2_{n-1,\alpha/2})$. The symbol $t_{n-1,1-\alpha/2}$ denotes the upper $1-\alpha/2$ quantile of the Student's-t distribution with $n-1$ degrees of freedom. A similar definition holds for $\chi^2_{n-1,1-\alpha/2}$.

Given a test of a hypothesis at the α-level of significance, the set of all values of the statistic that would lead to acceptance of the hypothesis has been defined by Kempthorne and Folks as a $(1-\alpha)100$ percent *consonance interval*, meaning that values of the parameter in such an interval are consonant with the data in the sample. The inversion of the standard tests of hypothesis leads to a consonance interval identical to the confidence interval constructed from the same data. The concept has not been widely used.

A *prediction interval* is a confidence interval in which a future observation takes the place of the unknown parameter, and in the long run a given fraction of such intervals will contain the "next" or a "future" observation generated from the population. Those intervals satisfy the probability statement $P(L < x_0 < U) = 1 - \alpha$, where x_0 represents the unknown future observation and L and U are based upon a sample. The quantity x_0 may also represent the mean of several future observations. Use of those intervals is standard.

A *tolerance interval* is an interval, constructed from a random sample, that includes a least at specified proportion of the population with a specified confidence level. Those intervals satisfy the double probability statement $P\{P[L \le X \le U] \ge \gamma\} = 1-\alpha$, meaning that in the long run a given fraction of those intervals will contain at least γ 100 percent of the population. For the normal distribution a tolerance interval is of the form $X \pm ks$, where k depends on α, γ, and n.

A *probability interval* is an interval (L,U) such that the probability that a random variable X is between L and U is $(1-\alpha)$. In mathematical terms if X is a random variable, then the $(1-\alpha)100$ percent probability interval (L,U) on X is such that $P(L \le X \le U) = 1-\alpha$. It may be stated that $(1-\alpha)100$ percent of the population of X is between L and U. In that case L and U are not usually random variables.

A *Bayesian interval* or *Bayesian confidence interval* is a probability interval on the posterior density. In other words, if $g(\theta|X)$ is the posterior density, the Bayesian interval is given by $L < \theta < U$ such that

$$P(L \le \theta \le U|X) = \int_L^U g(\theta|x)d\theta = 1-\alpha.$$

A *fiducial interval* is a probabilitylike interval formed around an unknown parameter using the parameter's fiducial distribution. Let T be a continuous sufficient statistic for θ and $F(t;\theta)$ the distribution function of T. If t is a realization of the random variable T, the *fiducial distribution* of θ, as defined by Fisher, is $G(\theta) = p$ where θ is the solution to the equation $F(t,\theta) = 1-p$. If θ increases as p increases, the fiducial interval is given by $L \leq \theta \leq U$ such that $G(L) - G(U) = 1-\alpha$. Under the correct conditions the likelihood function $L(\theta)$, scaled by $\int_{-\infty}^{\infty} L(\theta)d\theta$, may be called the *fiducial density*: In that case the fiducial interval is given by the L and U such that

$$\int_{L}^{U} L(\theta) \ d\theta \ / \int_{-\infty}^{\infty} L(\theta)d\theta = 1-\alpha.$$

Kendall and Stuart have stated that "the fiducial distribution is not a probability distribution in the sense of the frequency theory of probability. It may be regarded as a distribution of probability in the sense of degrees of belief. . . . " Fiducial intervals are understood by only a few and used by fewer still. In comparing confidence and fiducial intervals, Kendall and Stuart state that the confidence approach says that "we shall be right in about $(1-\alpha)100$ percent of the cases in the long run." However, the fiducial approach says that "(in some sense not defined)" we are $(1-\alpha)100$ percent sure of being right in this particular case." A *relative likelihood interval* is defined as those values of θ such that $L(\theta)/L(\hat{\theta}) \geq \alpha$, where $L(\theta)$ is the likelihood function for θ and $\hat{\theta}$ is the maximum likelihood estimator. Those intervals are rare in practice.

Any of the above methods may be extended to more than 1 parameter, population, or future observation, and the intervals generated are called *simultaneous intervals or regions*.

Hypothesis Testing

A *statistical hypothesis* is an assertion or a conjecture about the distribution of 1 or more random variables. If the hypothesis is specifically concerned with the values of 1 or more parameters, it is said to be *parametric*; otherwise it is *nonparametric*. If the assertion completely specifies the parameters of the distribution (e.g., $\theta = 5$), it is called a *simple hypothesis*, otherwise (e.g., $\theta \leq 5$) it is said to be a *composite hypothesis*. The hypothesis to be tested is the *null hypothesis*, denoted by H_0, and its negation is called the *alternative hypothesis*, denoted by H_A. The basic problem is to decide, on the basis of the outcome of an experiment, whether the null hypothesis is true. A *test of hypothesis* is a rule or procedure for deciding whether to *reject* the null hypothesis. *Rejection* is a decision that the observations are not favorable to the hypothesis. If H_0 is not rejected, it is *accepted* by default, i.e., for lack of evidence to the contrary. If H_0 is rejected, H_A is accepted.

A *nonrandomized test* procedure consists of dividing the sample space S (of possible outcomes of the experiment) into a region ω and its complement $S-\omega$. If the outcome s falls into region ω, H_0 is rejected; and if s falls into $S-\omega$, H_0 is accepted. The region that leads to rejection of the null hypothesis is called the *critical region* (*rejection region*), and its complement is called the *acceptance region*. In certain circumstances the procedure can be varied. A *randomized test* is one in which a statistic, $0 \leq \phi(x) \leq 1$, depending on the outcome s, is calculated. A Bernoulli trial (such as a coin toss) is then performed, which has $\phi(x)$ as a probability of success. If the Bernoulli trial results in a success, H_0 is rejected. (All the tests we discuss will be of the nonrandomized type).

When testing a hypothesis, the experimenter may make the correct decision or commit 1 of 2 errors: (a) reject the null hypothesis when it is true (referred to as an *error of the first kind* or a *Type I error*) or (b) accept the null hypothesis when it is false (called an *error of the second kind* or a *Type II error*). It is customary to assign a least upper bound to the probability of incorrectly rejecting the null hypothesis when it is true and to call the bound α the *level of significance* or the *size of the test* or *the size of the critical region*. There is no such thing as a *level of confidence* for a test of hypothesis. The probability β of making a Type II error is a function of the alternative hypothesis. If θ parameterizes the distributions in the alternative hypothesis, β is a function of θ and we call $\beta(\theta)$ the *operating characteristic function* and $1-\beta(\theta)$ the *power function*. Evaluated for a specific distribution in the class of alternatives, $1-\beta(\theta)$ gives the *power of the test* against that alternative, i.e., the probability of correctly rejecting the null hypothesis.

The choice of the level of significance is arbitrary, but it is desirable to make the risk of a Type I error small; hence it is conventional to limit the values of α to the set $\{0.10, 0.05, 0.025, 0.01, 0.005\}$. The benefit of that convention is that the number of tables needed is greatly reduced. In practice a test is carried out by calculating a test statistic (a function of the observations) whose distribution is known when the null hypothesis is true, formulating a critical region (a function of the significance level) and deciding whether the test statistic falls into the critical region. In applications it is good practice not only to tell whether the hypothesis was rejected but also to give a measure of the strength of the rejection. That is done by giving the *P-value*, the smallest significance level at which the hypothesis would be rejected.

For illustrative purposes, it may be useful to give an account here of the more common tests of hypotheses: (1) To test whether the mean μ of a normal distribution is equal to a specified value, μ_0, when the standard deviation σ is known, we calculate the statistic $U = (\overline{X}-\mu_0)/(\sigma/\sqrt{n})$. U has a $N(0,1)$ distribution under the null hypothesis, $H_0 : \mu = \mu_0$. If $|U|$ exceeds the $(1-\alpha/$

2)100th percentile of the $N(0,1)$ distribution, we reject the null hypothesis and decide that μ differs from μ_0. That is the *one-sample normal test*. (2) To test the same hypothesis as above when σ is unknown (and has to be estimated from the sample), we calculate $U = (\overline{X} - \mu_0)/(S/\sqrt{n})$, where S is the sample standard deviation. Under the null hypothesis, U has a student's t distribution with $(n\text{-}1)$ *degrees of freedom*; hence we reject H_0 at the α level of significance if $|U|$ exceeds the $(1 - \alpha/2)$ 100th percentile of the Student's t distribution. This is called a one-*sample t-test*. (3) To test whether the means of 2 normal populations are equal, assuming they have equal variances, we calculate the statistic $U = (\overline{X} - \overline{Y})/S_p(1/n_x + 1/n_y)^{1/2}$ where S_p^2 is the *pooled variance* $[(n_x - 1)S_x^2 + (n_y - 1)S_y^2]/(n_x + n_y - 2)$. Under H_0, U has a Student's t distribution with $n_x + n_y - 2$ degrees of freedom; hence we reject $H_0 : \mu_x = \mu y$ if $|U|$ exceeds the $(1 - \alpha/2)$100th percentile of the Student's t distribution. Again, the level of significance is α. It is a *2-sample t-test*. If the 2 variances cannot be assumed to be equal, we use $(S_x^2/n_x + S_y^2/n_y)^{1/2}$ in the denominator and calculate the approximate degrees of freedom by *Satterthwaite's rule*. The rule states that the degrees of freedom for a linear combination $Y = \Sigma a_i s_i^2$ of independent mean squares are $Y^2/\Sigma(a_i s_i^2)/fi$ where f_i is the degrees of freedom of s_i^2. The degrees of freedom are those of the chi-square variate that match the first 2 moments of Y. (4) When the observations from 2 normal samples are naturally paired, we take the differences d_i and do a one-sample t-test on them. It is called a *paired differences t-test* or a *paired t-test*.

If the 2 variances can be assumed equal, we have *homoscedasticity*; otherwise we have *heteroscedasticity*. The problem of unequal variances is referred to as the Behrens-Fisher problem. (5) To test the hypothesis that the standard deviation σ of a normal population is equal to a specified value σ_0, calculate the statistic $U = (n - 1)S^2/\sigma_0^2$. Under the null hypothesis $H_0 : \sigma = \sigma_0$, U has a chi-square distribution with $(n - 1)$ degrees of freedom. Using a level of significance of α, reject the null hypothesis in favor of the hypothesis $\sigma > \sigma_0$ if U exceeds the $1 - \alpha$ percentile of the chi-square distribution. (6) To test whether the standard deviations of 2 normal populations are equal, calculate $U = S_x^2/S_y^2$, where S_x^2 is the largest variance, and reject the null hypothesis that $\sigma_x = \sigma_y$ in favor of $\sigma_x > \sigma_y$ if U exceeds the $1 - \alpha$ percentile of the F-distribution with $(n_x - 1)$ and $(n_y - 1)$ degrees of freedom. The first 3 tests are two-sided tests while the last 2 are one-sided.

When we wish to test the equality of several means from normal populations, we use the *analysis of variance*. To test the equality of several variances, a number of tests are available, such as *Bartlett's test*, which uses as a statistic B/C where $B = -\Sigma f_i \ln(S_i^2/S^2)$, $C = 1 + (\Sigma(1/f_i - 1/f))/3(K - 1)$, where the variance S_i^2 has f_i degrees of freedom, $f = \Sigma f_i$ and $S^2 = \Sigma f_i S_i^2/f$; *Cochran's test*, which uses as a statistic the largest variance over the sum of the variances; and *Hartley's max-F test*, which uses as a statistic the largest variance in the set divided by the smallest. Tests of means are *robust* (insensitive) to the normality assumption, while tests of variance are notoriously sensitive to normality.

All of the above tests are *parametric tests*, which depend upon the distribution assumed. In most cases there is a corresponding *nonparametric* or *distribution-free* test. They are applicable to a much wider class of situations but lose some power in the process.

The nonparametric tests are based on *ranks*. (In a sample of size n, the smallest observation has rank 1, the next largest rank 2, and so on to the largest, which has rank n.) For the *sign test* we classify the i-th pair as $+$ if $x_i < y_i$, $-$ if $x_i > y_i$ and a "tie" if $x_i = y_i$. The null hypothesis is that $P(X_i < Y_i) = 1/2$. Let T be the number of $+$'s and n the number of observations not tied. If T is less than the lower tabled value or larger than the upper $1 - \alpha/2$ tabled value, we reject H_0 at the α-level of significance in favor of one sign being more prevalent. That is the oldest, quickest, and most useful of the nonparametric tests and can be applied in dozens of situations. By using the magnitude as well as the sign of the difference, we obtain the more powerful *Wilcoxon Signed Rank* Test. Again disregarding ties, we rank the absolute differences $D_i = |Y_i - X_i|$. For observations with equal values of D_i, we assign the average rank. Let T be the sum of the ranks for which $Y_i > X_i$. If T is larger than the tabled values, we reject the hypothesis that the common median of the D's is zero and decide that the X's tend to be larger than the Y's or vice versa. Either of the above will serve as a test of equal medians for the 2 populations. For k populations, we construct the *Median Test* by making a $2 \times k$ table. We classify a random sample of n_i observations from the i-th population as above or below the median and apply a chi-square test of independence to this contingency table.

We considered above a set of naturally paired observations (X_1, Y_1), (X_2, Y_2), . . . such as "before" and "after" data. In a different setting we are given 2 samples of different sizes: $X_1, \ldots X_n$ and $Y_1 \ldots Y_m$. We wish to test the hypothesis that they have the same cdf. More particularly, we assume that the only difference in their cdf is the median so that the null hypothesis is that they have the same median. We combine all $n + m$ points and rank them. Let S be the sum of the ranks of the X's. The minimum value for S is $n(n+1)/2$. The test statistic is $T = S - n(n + 1)/2$. If T exceeds the tabulated critical value, we decide that the X's tend to be larger than the Y's. That is the *Mann-Whitney* or *Rank-Sum* or *Wilcoxon two-sample test*. If there are k such samples rather than 2, we proceed in the same way, letting R_i be the sum of the ranks of observations in the i-th sample, with $N = \Sigma n_i$. The test statistic is $T = 12(N(N+1))^{-1} \Sigma(R_i^2/n_i) - 3(N + 1)$. If T exceeds the critical value, the population medians are not equal. That generalization of the Mann-Whitney test is called the *Kruskal-Wallis Test* for k samples. The *Bell-Doksum Test* is a modification of the Kruskal-Wallis test that uses tabled values of the normal order statistics instead of ranks. The Wilcoxon Signed Rank Test is analogous to the Paired t-test; the Mann-Whitney test is analogous to the two-sample t-test, and the Kurskal-Wallis test is the analogue of a one-way ANOVA.

For a sample of size n and outcomes that are either "success" of "failure" (Bernoulli trials), we are interested in whether the probability p of success is equal to a specified value. In a plant-breeding experiment, for example, we may be interested in whether the proportion of tall plants is 3/4. From a binomial table with arguments p and n, we find t_1 such that P (less than t_1 successes) $= \alpha_1$, and t_2 such that P (more than t_2 successes) $= \alpha_2$ where $\alpha = \alpha_1 + \alpha_2$. That is the *binomial test*. We cannot set α at exactly .05 for a discrete distribution. We reject the null hypothesis if the observed number of successes in n trials is less than t_1 or greater than t_2. The sign test is a binomial test with $p = 1/2$. We can use it to test whether the p-th quantile is equal to a specified value. It is then called the *quantile test*.

Crucial to all of the above tests is the assumption that the observations are *i.i.d.* A possible form of nonindependence is correlation between consecutive observations, called *serial correlation* or *autocorrelation*, meaning that observations a distance of k units apart are correlated. A test of independence versus serial correlation is the *mean square successive difference test*. We calculate the sample variance S^2 and $d^2 = \Sigma(x_{i+1} - x_i)^2/(n-1)$. The test statistic is $T = (d^2/2S^2 - 1)[(n-2)(n^2-1)]$ and is approximately $N(0,1)$. If T is *less* than the lower critical value of the unit normal, independence is rejected in favor of serial correlation. A second test, the *runs test*, is nonparametric. Observations from a continuous variate are classified as A (above the median) or B (below the median). We ignore observations equal to the median. The number of runs R is 1 plus the number of changes (from A to B or B to A). Let m be the number of observations above the median. Then the statistic $T = [(R + 1/2) - (1 + m)]/[m(m-1)/(2m-1)]^{1/2}$ is approximately $N(0,1)$ and is used to test the hypothesis of independence. A two-sample runs test is constructed by labeling observations from one population as A and those from the other as B. We then rank all the observations and calculate the number of runs R. We reject the hypothesis of equal distribution functions if R exceeds the tabled critical values.

For a test of a simple hypothesis against a simple alternative, a *best critical region* is one whose power is no smaller than any other region of the same size, and a test based on a best critical region is a *most powerful test* or *best test*. A most powerful test of size α is one of size α that has greater power than any other test of the same size. Most powerful tests always exist for simple null and alternative hypothesis, and in such cases the construction of a best critical region of size α is straightforward. Let $f(x;\theta)$ be the probability density function of the observations, and let $X_1, \ldots, X_n$ denote a random sample from that density. Then the joint pdf of $X_1, \ldots, X_n$ is $L(\theta; x_1, \ldots, x_n) = f(x_1;\theta) f(x_2;\theta) \ldots f(x_n;\theta)$, which is called the *likelihood function*. We wish to test the null hypothesis $H_0:\theta = \theta_0$ against the simple alternative $H_A:\theta = \theta_1$. Let L_0 and L_1 be the likelihood function evaluated at $\theta = \theta_0$ and $\theta = \theta_1$ respectively. The *likelihood ratio* is defined as $\upsilon = L_0/L_1$. Let k be some positive number (which will depend upon α). The *Neyman-Pearson Lemma* states that C is a best critical region of size α if (a) it is a critical region of size α, (b) $\lambda \leq k$ for all x in C, and (c) $\lambda \geq k$ for x not in C. The test statistic will then be λ (or some monotone function of λ), and k will depend upon the distribution of λ under the null hypothesis. A test based upon λ is a most powerful test of size α and is called a *likelihood ratio test*; the corresponding region is the *likelihood ratio critical region*. Kendall has said that "nearly all tests now in use for testing parametric hypothesis are likelihood ratio tests, . . . but cases exist in which the likelihood ratio is not only unsatisfactory, but worse than useless."

Best tests depend, in general, upon the simple alternative H_A. In some situations the critical region is independent of the particular alternative, i.e., there exists 1 best critical region that is best for every distribution in the class of distributions defined by the composite alternative hypothesis; it has the largest power among all tests of the same size of *all* alternative values of the parameter θ. In that case both the critical region and the test are described as *uniformly most powerful*. When a uniformly most powerful test does not exist, it may be possible to find a region close—in some sense—to the null hypothesis and a test T whose power is greater than that of any other test of the same size. The test T is then said to be *locally most powerful*. A family of densities $f(x;\theta)$, where θ lies in some interval, has a *monotone likelihood ratio* if there exists a function $t(x_1, \ldots, x_n)$ such that the likelihood ratio $L(\theta_1)/L(\theta_2)$, for every $\theta_1 < \theta_2$, is a monotone function of t. A uniformly most powerful test exists for one-sided hypothesis if the density sampled from has a monotone likelihood ratio in some statistic. Even more important, uniformly most powerful tests are always based upon a sufficient statistic if one exists.

A desirable property of a test is *unbiasedness*. A test is *unbiased* if the probability of rejecting H_0 when H_0 is false is always greater than or equal to the probability of rejecting H_0 when H_0 is true. Tests are also described as *one-tailed* or *two-tailed* depending upon whether the rejection region lies in only one or in both tails of the test statistic's distribution. If the alternative hypothesis is such that the inequality proceeds in only one direction, the test is said to be *one-sided*. If the inequality may go in either direction, the test is *two-sided*. A *conservative* test is one for which the actual level of significance is smaller than the stated level.

A composite hypothesis leaves some parameters unspecified. The size of the critical region under the null hypothesis usually depends upon the unspecified parameters. Consider the boundary of the 2 parameter sets defined by the null and alternative hypothesis; if the critical region has the same size α for all parameter values on this boundary, it is said to be *similar to the sample space*, and a test based upon such a region is a *similar size α test*. A test of a hypothesis H_0 against a class of alternatives H_A is said to be *consistent*, if, when any member of H_A holds, the probability of rejecting H_0 tends to 1 as the sample size gets large. A test that minimizes the amount by which the actual power falls short of the maximum power attainable is called a *most stringent test*.

Another approach to hypothesis testing is based on decision theory. Basic to the theory is the assumption of a set of possible *states of nature*, also referred to as the *parameter space*. The parameter space characterizes a family of distributions $f(x;\theta)$, with each value of the parameter representing a different distribution. There is also the *action space*, which is the set of actions or decisions $\{d_i\}$ available to the decision maker. Finally there is a function of the parameter θ and the action d called the *loss function $L(\theta,d)$*. Nature assumes a point θ in the parameter space, and the decision maker, without being informed of the state of nature, chooses an action or makes a decision $d(x)$, which is based upon the outcome x of an experiment. As a consequence of those 2 choices, the loss is an amount $L[\theta,d(X)]$, called the *risk function $R(\theta,d)$*, and represents the average loss to the statistician, and the function $d(x)$ (which maps the sample space in the action space) is called a *nonrandomized decision rule* or *decision function* provided that $R(\theta,d)$ exists and is finite for all θ in the parameter space. A *randomized decision rule* allows the choice of action to be made by some random mechanism, while the decision maker makes some (educated) guess of the probabilities (the probability density function) upon which the random mechanism operates. The fundamental problem of decision theory is that of choosing (selecting) a decision or a course of action. If the decision maker is willing to describe his degree of belief concerning the states of nature by assigning a probability to each state of nature before taking a sample, we refer to that set of probabilities as the *prior probability distribution*. It is reasonable to measure the consequence of a procedure d by calculating the expected risk (with respect to the prior probability distribution τ) and calling it the *Bayes risk* of a decision rule d with respect to a prior distribution τ. A *Bayes decision rule* is the one that has smallest Bayes risk. The *Bayes principle* calls for use of the Bayes decision rule. Since both the outcome X of the experiment and the parameter θ are being tested as random variables, we can speak of their joint distribution and of the conditional distribution of θ given X, which is called the *posterior distribution* of θ. A Bayes decision rule minimizes the posterior conditional expected loss, given the observations. A second criterion for choosing a decision rule is the *minimax principle*: for every action, the decision maker finds the maximum loss that can be incurred (as a function of the possible states of nature) and minimizes the maximum loss. A test using that principle is a *minimax test*.

A decision rule d_1 is said to be *as good as* rule d_2 if the risk function $R(\theta, d_1) < R(\theta, d_2)$ for all θ in the parameter space. Rule d_1 is *better than* rule d_2 if it is as good as rule d_2 and $R(\theta, d_1) \leq R(\theta, d_2)$ for at least 1 value of θ. A rule d is *admissible* if there exists no rule better than d. A desirable property of a decision rule is that it be *invariant*. Let $P(\theta)$ be a family of distributions with θ in a parameter space Θ. *The family of distributions is invariant* under a group G of transformations if for every g in G and θ in Θ there exists a unique θ' in Θ such that the distribution of $g(X)$ is $P(\theta')$ when the distribution of X is $P(\theta)$. Given an invariant family of distributions and the value of θ' determined by a particular value of g and θ, *a decision problem is invariant* if for every g in G and every a in the action space A, there exists a unique a' in A such that the loss function is invariant, i.e., $L(\theta, a) = L(\theta', a')$. The a' uniquely determined by g and a is denoted by $\bar{g}(a)$. Given the θ' and a' uniquely determined by an invariant decision problem, and a nonrandomized decision rule $d(x)$, *the decision rule is invariant* if for every g in G and every x in the sample space, $d[g(x)] = \bar{g}[d(x)]$. A class C of decision rules is *complete (essentially complete)* if C is a subset of D and if, given a rule d in D but not in C, there exists a rule d_0 in C that is better than (as good as) d. The class C of decision rules is *minimal complete* if C is essentially complete and no proper subclass of C is essentially complete. Thus there is no need to look outside an (essentially complete) complete class for a decision rule; one can do (just as well) or better inside the class.

References

A good elementary text that is easy reading is Dixon, W. J., and Massey, F. J. 1983. *Introduction to Statistical Analysis*. 4th ed. Following this I recommend Brownlee, K. A. 1965. *Statistical Theory and Methodology in Science and Engineering*, 2nd ed. New York: Wiley, or Bowker, A. H., and Lieberman, G. J. 1972. *Engineering Statistics*. 2nd ed. Englewood Cliffs, N.J.: Prentice-Hall. A good step-by-step text in simple linear regression and hypothesis testing is Natrella, M. G. 1963. *Experimental Statistics*, NBS Handbook 91, Govt. Printing Office, Washington, D.C. A book with valuable information on less-standard techniques is Hahn, G. J., and Shapiro, S. S. 1967. *Statistical Models in Engineering*, New York: Wiley and Sons. The 2 classics in mathematical statistics are Mood, A. M., Graybill, F. A., and Boes, D. C. 1974. *Introduction to the Theory of Statistics*. 3rd ed. New York: McGraw-Hill, and Hogg, R. V., and Craig, A. T. 1978. *Introduction to*

Mathematical Statistics, 4th ed. New York: MacMillan. In the area of Bayesian estimation, a thorough treatment and link to classical statistics is given in Box, G.E.P., and Tiao, G. 1973. *Bayesian Inference in Statistical Analysis*, Reading, Mass.: Addison-Wesley. The very important topic of nonparametric statistics is well covered in Conover, W. J. 1980. *Practical Nonparametric Statistics*. 2nd ed. New York: Wiley and Sons.

5

Regression

A common problem is that of estimating a *linear relationship* between 2 variables, x and y. If an electrician charges a flat fee β_0 plus a fixed amount β_1 per outlet when he wires a house, the relationship can be expressed as $y = \beta_0 + \beta_1 x$, where y is his total fee and x is the number of outlets. There is no error in the fee because we can count the number of outlets. For a given number of outlets the fee is invariably the same. In that case both variables are *mathematical* or *nonrandom variables*, i.e., they do not have distributions. When y is plotted against x (for any fixed value of β_0 and β_1), the points fall on a straight line with *slope* β_1 and *intercept* β_0 and we say that the relationship is *linear*. The quantities β_0 and β_1 are *parameters*; they are fixed constants in any given situation but may vary from one situation to another. If β_0 and β_1 were unknown, they could be determined exactly if one had at hand 2 or more distinct values of x and y.

A second case of major importance is the one in which y is a random variable and x is a fixed mathematical variable, measured without error. For example, let x be the amount of phosphate added to a plot of ground of a certain size and let y be the yield of corn from that plot of ground. For any given x, the yield, y, varies with the soil type, moisture, sunshine, variety of seed, etc., so that there is a distribution of y-values for a given x. Since

47

we cannot predict the exact outcome with certainty, we content ourselves with estimating the *average responses* that we assume is linear with x. The yield for the i-th plot may thus be represented as $y_i = \beta_0 + \beta_1 x_i + \epsilon_i$, where ϵ_i is a random error. The average response is $E(y_i) = \beta_0 + \beta_1 x_i$. Since we control x and observe the result y, we say that y is the *dependent variable* (depends on x) and that x is the *independent variable*.

In a third case we sample n individuals and observe 2 characteristics, x and y, on each individual. We arbitrarily select 1 of the variables, which we wish to predict, and "call" it the dependent variable. We assume a linear relationship between x and y. In that case x *and* y are random variables with a bivariate distribution. Galton selected a group of men and measured the height y of each man and the height x of his father. After plotting the points, he observed that, on the average, sons of tall fathers were not so tall as their fathers and sons of short fathers were not so short as their fathers. He called that tendency a *regression* toward the mean, and the term caught on and is now applied loosely to the estimation of almost any type of relationship in which 1 or more of the variables is random. Statisticians restrict the word to situations in which the dependent variable is random and the independent variables are fixed, mathematical variables.

We shall now be more precise in our definitions. *Regression analysis* is applicable in situations in which the expected value of a random variable Y depends upon the values of other variables $X_1, X_2 \ldots , X_p$, which are called *independent variables, regression variables, predictors, regressors, carriers,* or *exogenous variables*. Y is called the *dependent variable, response variable, predictand, regressand,* or *endogenous variable*. The *regression model, regression function, regression equation, regression surface,* or *response surface* expressing that relationship may be written:

$$E(Y|X_1, X_2, \ldots , X_p) = f(X_1, X_2 \ldots , X_p; \beta_0, \beta_1, \ldots , \beta_p)$$

where the β_i are unknown constants or *parameters*, which we are required to estimate. The X_j are nonrandom, mathematical variables. For the *i-th case* or *i-th data point* $(Y_i, Y_{i1}, \ldots , X_{ip})$, the X_{ij} are known, fixed constants, measured without error or with negligible error. If the model is linear in the parameters, it is a *linear model* as opposed to a *nonlinear* or *curvilinear model*. (A model is linear in the parameter β_i only if $\partial Y/\partial \beta_i$ is not a function of β_i). Thus $y = \beta_0 \sin x$ is linear, but $y = \sin \beta_0 x$ is nonlinear.

In scalar notation we represent the i-th data point of a linear model as

$$Y_i = \beta_0 + \beta_1 X_{i1} + \ldots + \beta_p X_{ip} + \epsilon_i.$$

The expected value, $E(Y_i) = \beta_0 + \beta_1 X_i^1 + \ldots + \beta_p X_{ip}$, is interpretable as a *regression hyperplane*. If $p = 1$, the expected value is a straight line and the model is a *simple linear regression* model; if $p > 1$, it is a *multiple linear regression* model. β_0 is the *intercept*, and $\beta_1, \ldots, \beta_p$ are called the *partial regression coefficients*; they are slopes of the regression plane in various directions. If $\beta_0 = 0$, we have a *regression through the origin*. The linear model in which $X_{ij} = X_i^j$ is a *polynomial regression function* or *polynomial model*. The *order* of such a model is the highest value of j present in the model.

Regression in a linear model may be *qualitative* in nature, i.e., the independent variables assume only values of 1 or 0 and represent a classification; they are called *dummy variables, indicator variables,* or *carriers*. If all the regressors are qualitative, the linear model is an *analysis of variance model*; if only part are qualitative, the model is an *analysis of covariance model*. The quantitative independent variables in the analysis of covariance model are called *covariates* or *concomitant variables*. When an independent variable is the product of 2 other regressors, it is said to represent an *interaction term* in the model. There may also be interaction terms that are not products. In an analysis of variance model, for example, any term except the main effects is called an interaction term; it need not be any specific function of the main effects. A model is *additive* if it contains no interaction terms and *nonadditive* if it does.

For the sake of conciseness the *linear model* may be written in matrix notation as $\mathbf{Y} = \mathbf{Y}\boldsymbol{\beta} + \boldsymbol{\epsilon}$, where $\mathbf{Y}$ is the n $\times$ 1 vector of observations, $\mathbf{X}$ is an $nx(p + 1)$ matrix of known constants, $\boldsymbol{\beta}$ is the $(p + 1) \times 1$ vector of parameters and $\boldsymbol{\epsilon}$ then $n \times 1$ vector of random errors. The matrix $\mathbf{X}$ is the *design matrix, model matrix, data matrix,* or *incidence matrix*, depending on the context. A complete description of the model depends on the assumptions made about the vector $\boldsymbol{\epsilon}$. It is usually assumed that $E(\boldsymbol{\epsilon}) = 0$ and $\text{Var}(\boldsymbol{\epsilon}) = \sigma^2 \boldsymbol{\Sigma}$, where $\boldsymbol{\Sigma}$ is a positive definite matrix and σ^2 is a positive constant. In addition, the assumption that $\boldsymbol{\epsilon}$ is multivariate normal is usually required for hypothesis testing.

Hypothesis testing in the linear model is frequently a matter of deciding whether or not to include certain terms in the model. That is done by fitting a *full* or *unconstrained model* and then fitting a *reduced, constrained, restricted,* or *subset model* in which the terms under consideration are dropped, i.e., their coefficients are set equal to zero. The *reduction in the sum of squares* of residuals (deviations from the fitted model) is the difference in sums of squares (for the constrained model minus the unconstrained model), sometimes called the *extra sum of squares*. That quantity, divided by the difference in the degrees of freedom for the 2 residual sums of squares, is divided by the residual mean square for the full model and compared with tabulated values of the F distribution. If the statistic exceeds the tabulated value, the terms in question are retained. That procedure is used to build a model in stages, called a *stepwise regression* procedure. If we begin with the full model and drop out terms 1 at a time, we have a *backward elimination procedure*, while if we begin with β_0 and add terms 1 at a time, we have a *forward selection procedure*. If a model is selected from regressions on all possible subsets of the regressors, we have a *best subsets regression* procedure.

A number of criteria are available for constructing estimators of β and σ^2. The most common criterion, *least squares*, specifies that the estimator, $\hat{\beta}$, of β be chosen to minimize $(\mathbf{Y} - \mathbf{X}\hat{\beta})' \, \mathbf{\Sigma}^{-1} \, (\mathbf{Y} - \mathbf{X}\hat{\beta})$, which, for $\mathbf{\Sigma} = \mathbf{I}$ (the identity matrix), is the sum of squares of deviations between the observed and expected values: $\Sigma(y_i - \beta_0 - \beta_1 X_{i1} - \ldots - \beta_p X_{ip})^2$.

The resulting estimator is

$$\hat{\beta} = (\mathbf{X}'\mathbf{\Sigma}^{-1}X)^{-1} \, X'\mathbf{\Sigma}^{-1}Y$$

which may be found as the solution to the *normal equations*

$$(X'\mathbf{\Sigma}^{-1}X)\hat{\beta} = X'\mathbf{\Sigma}^{-1}Y.$$

For simple linear regression, $\hat{\beta}_1 = (n\Sigma xy - \Sigma x \Sigma y)/(n\Sigma x^2 - (\Sigma x)^2)$ and $\beta_0 = \bar{y} - \hat{\beta}_1 \bar{x}$. The variance about that line is $s_{y \cdot x}^2 = (n-1)s_y^2 - \hat{\beta}_1^2 s_x^2)/(n-2)$, and the variance of $\hat{\beta}_1$ is $s_\beta^2 = s_{y \cdot x}^2/(n-1)s_x^2$. A test of whether the slope is zero is made by calculating $\hat{\beta}_1/s_\beta$. If that quantity is less in absolute value than $t_{n-2, 1-\alpha/2}$, there is no predictive value in the x's (the slope is not different from zero). In *calibration* we use n pairs of data points (x_i, y_i) to establish a *calibration line* or a *calibration curve* $y = \alpha + \beta x$. The n values of x are standards or known values. The y-values are measurements of the standard values by the instrument we wish to calibrate. The line is estimated by the usual least squares fitting procedure. The usual problem in regression is to predict a y-value from a given x-value, but the *calibration problem* is the use of the regression line in reverse: Given a new measurement y_{n+1} on the instrument, we wish to obtain the "true" or "known" value, x_{n+1}. Furthermore, we would like to place an interval around the x-value, and there are several ways of doing that.

If $\mathbf{\Sigma}$ is the identity matrix, $\hat{\beta}$ is called the *ordinary least squares* estimator of β and is the *best linear unbiased estimator, (BLUE)* when $E(\mathbf{e}) = 0$ and

Var($\mathbf{e}$) = $\sigma^2\mathbf{I}$. That result is known as the *Gauss-Markov Theorem*. If $\mathbf{\Sigma}$ is a diagonal matrix not equal to the identity matrix, $\hat{\boldsymbol{\beta}}$ is called a *weighted least squares estimator* of $\boldsymbol{\beta}$ and the diagonal elements of $\mathbf{\Sigma}$ are called weights or case *weights*. Otherwise ($\mathbf{\Sigma}$ not diagonal), $\hat{\boldsymbol{\beta}}$ is called the *generalized least squares* estimator of $\boldsymbol{\beta}$. If $\mathbf{e}$ follows a multivariate normal distribution, then the distribution of $\hat{\boldsymbol{\beta}}$ is normal with mean $\boldsymbol{\beta}$ and *covariance* or *dispersion matrix*, Var($\hat{\boldsymbol{\beta}}$) = $\sigma^2(\mathbf{X'\Sigma}^{-1}\mathbf{X})^{-1}$.

Since $\mathbf{\Sigma}$ is assumed known, the linear model with $\mathbf{\Sigma} \neq \mathbf{I}$ may be transformed to an equivalent model by writing

$$\mathbf{CY} = \mathbf{C\,X\,\boldsymbol{\beta}} + \mathbf{Ce}$$

or

$$\mathbf{Z} = \mathbf{W\,\boldsymbol{\beta}} + \boldsymbol{\delta}$$

where $\mathbf{\Sigma}^{-1} = \mathbf{C\,C'}$. In the transformed model, $E(\boldsymbol{\delta}) = 0$ and Var($\boldsymbol{\delta}$) = $\sigma^2\mathbf{I}$, and, therefore, ordinary least squares is applicable. In the following, without loss of generality, only ordinary least squares is considered.

The *ordinary least squares estimator* of $\boldsymbol{\beta}$ is, from the above,

$$\hat{\boldsymbol{\beta}} = (\mathbf{X'X})^{-1}\mathbf{X'Y}$$

The matrix $\mathbf{X'X}$ is called the *information matrix* or *cross-product matrix*. If the carriers are centered at zero by subtracting means, then $\mathbf{X'X}$ is called the *corrected cross-product matrix*. The *standard errors of the estimates* of the individual components of $\hat{\boldsymbol{\beta}}$ are the square roots of the diagonal elements of Var ($\hat{\boldsymbol{\beta}}$) = $\sigma^2(\mathbf{X'X})^{-1}$.

The *fitted values* or *predicted values*, $\hat{y}_i$, are the elements of

$$\hat{\mathbf{Y}} = (\hat{y}_i) = \mathbf{X}\hat{\boldsymbol{\beta}} = \mathbf{X}(\mathbf{X'X})^{-1}\mathbf{X'Y}$$

and Var($\hat{\mathbf{Y}}$) = $\sigma^2\mathbf{X}(\mathbf{X'X})^{-1}\mathbf{X'}$. The symmetric, independent, rank p matrix, $\mathbf{X}(\mathbf{X'X})^{-1}\mathbf{X'}$ is called the *hat matrix*. The *ordinary residuals*, r_i, are

$$\begin{aligned}\mathbf{R} = (r_i) &= \mathbf{Y} - \hat{\mathbf{Y}} \\ &= (\mathbf{I} - \mathbf{X}(\mathbf{X'X})^{-1}\mathbf{X'})\mathbf{Y}.\end{aligned}$$

In other words, an ordinary residual is the difference between the observed and fitted value. The covariance matrix of the vector of ordinary residuals is $\text{Var}(\mathbf{R}) = (\mathbf{I} - \mathbf{X}(\mathbf{X}'\mathbf{X})^{-1}\mathbf{X}')\,\sigma^2$.

The sum of squares

$$\mathbf{R}'\mathbf{R} = \sum_{i=1}^{n} r_i^2$$

is called the *sums of squares for error (SSE)* or *error sums of squares (ESS)* and may be used to construct an unbiased estimator, $\hat{\sigma}^2$, of σ^2

$$\hat{\sigma}^2 = \text{SSE}/(n-p-1)$$

with $n-p-l$ degrees of freedom. That estimate is often termed the *estimated mean square* or the *residual mean square*. Alternatively the residual sums of squares may be written as

$$\mathbf{R}'\mathbf{R} = \mathbf{Y}'\mathbf{Y} - \hat{\boldsymbol{\beta}}'\,\mathbf{X}'\,\mathbf{Y}$$

where $\mathbf{Y}'\mathbf{Y}$ is the total sums of squares (TSS) and $\hat{\boldsymbol{\beta}}'\,\mathbf{X}'\,\mathbf{Y}$ is the *regression sums of squares (RSS)*.

The *proportion of variability explained* by the carriers, is called R^2 or *Multiple R^2*, while R is the *multiple correlation coefficient*, since it is the correlation between $\mathbf{Y}$ and $\hat{Y}$ and defined by

$$R^2 = 1 - \text{ESS}/(\mathbf{Y}'\mathbf{Y} - n\overline{\mathbf{Y}}^2).$$

That is widely viewed as a measure of the strength of the relationship between the dependent variable Y and the independent variables. An alternate version, $\overline{R}^2$, is defined as

$$\overline{R}^2 = R^2 - (1 - R^2)\left(\frac{p}{n-p-1}\right)$$

and is terms R^2 *adjusted for degrees of freedom*.

Hypothesis Testing is carried out by partitioning $\mathbf{X}$ and $\boldsymbol{\beta}$ as $\mathbf{X} = (\mathbf{X}_1,\mathbf{X}_2)$ and $\mathbf{B}' = (\mathbf{B}'_1, \mathbf{B}'_2)$, where $\boldsymbol{\beta}_1$ is $q \times 1$. With that partition, the linear model may be written as

$$\mathbf{Y} = \mathbf{X}_1\boldsymbol{\beta}_1 + \mathbf{X}_2\boldsymbol{\beta}_2 + \mathbf{e}$$

A test of the hypothesis $H_0:\boldsymbol{\beta}_2 = \mathbf{0}$ vs. $H_A:\boldsymbol{\beta}_2 \neq \mathbf{0}$ is a comparison of the *full model* with the *subset model*

$$Y = X_1\beta_1 + e_1.$$

Let ESS_1 and ESS_2 denote the error sums of squares under the full model and subset model, respectively. The *F-statistic* for the hypothesis $H:\beta_2 = 0$ is

$$\frac{(ESS_2 - ESS_1) \div (p + 1 - q)}{SSE_1 \div (n - p - 1)}$$

which has an *F*-distribution with $p + 1 - q$ and $n - p - 1$ degrees of freedom under the null hypothesis. The difference $ESS_2 - ESS_1$ is termed the *sums of squares for regression of β_2 adjusted for β_1* and may be written as

$$EES_2 - ESS_1 = RSS_1 - RSS_2$$

where RSS_2 and RSS_1 are the regression sums of squares for the subset model and full model, respectively. If β_1 contains only the intercept β_0, then $RSS_2 = n\bar{y}^2$ and $ESS_2 = \Sigma(y_i - \bar{y})^2$. When β_2 contains a single coefficient, say β_j, the *F*-statistic reduces to $\hat{\beta}_j^2/Var(\hat{\beta}_j)$.

The calculations necessary to obtain the *F*-statistic and the residual mean square are usually organized in an *Analysis of Variance Table (ANOVA Table)*: The *F*-statistic is the ratio of the 2 mean squares given in the last column of the ANOVA Table.

In some cases of regression, one point may, if left out, change the regression coefficients drastically, and a procedure for detecting such points is necessary. A *functional* is a correspondence between a class of functions and a set of real numbers. An example is $T = \int_a^b \alpha(x)y(x)dx$, where the $y(x)$ are the functions that map onto real numbers. In particular, if we let $\alpha(x) = x$,

Source of Variation	Degrees of Freedom	Sums of Squares	Mean Squares
Regression from full model, (β_1, β_2)	$p + 1$	RSS_1	
Regression from Subset model, (β_1)	q	RSS_2	
β_2 adjusted for β_1	$p + 1 - q$	$RSS_1 - RSS_2$	$\frac{RSS_1 - RSS_2}{p + 1 - q}$
Residual or Error	$n - p - 1$	ESS_1	$\frac{ESS_1}{n - p - 1}$
Total	n	TSS	

and $y(x)$ be a pdf with cdf $= F(x)$, $a = -\infty$, and $b = \infty$, then $T(F)$ is the mean of the distribution F. If we let δ_x be a pdf with all the probability at the point x, then a mixture of the distribution F and δ_x can be written as $(1-\epsilon)F + \epsilon\delta_x$ where ϵ is some number between 0 and 1. The *influence curve* of T at F is defined to be the limit as $\epsilon \to O$ of $T[(1-\epsilon)F + \epsilon\delta_x]/\epsilon$. If we let μ be the mean of F, the influence curve for the sample mean is $x - \mu$ and for the variance $(x-\mu)^2 - \sigma^2$. The influence curve is essentially the derivative of the estimator T. The influence curve, when plotted as a function of x for an estimator, indicates how the estimator changes if a new observation is thrown in at the point x, and it may be used to compare estimators. The *empirical influence curve* is obtained by substituting the empirical cdf for F in the equation definition of the influence curve. The *sample influence curve* is found by substituting $\epsilon = -1/(n-1)$ in the empirical influence curve. It is computed from the data and represents the change in a statistic when 1 observation is deleted.

In a regression problem we let $\boldsymbol{\beta}_{(i)}$ denote the vector of coefficients with the i-th case deleted. We then define $D_i = (\hat{\boldsymbol{\beta}}_{(i)} - \hat{\boldsymbol{\beta}})' \mathbf{X'X} (\hat{\boldsymbol{\beta}}_{(i)} - \hat{\boldsymbol{\beta}})/ps^2 = t_i V(Y_i)/_p V(R_i)$, where t_i is the i-th studentized residual, $V(R_i)$ is the variance of the i-th residual, and $V(Y_i)$ is the variance of the i-th predicted value. D_i is called *Cook's distance* and is used as a measure of influence (values around 3 are quite influential). Once detected, influence points should be considered carefully.

If the i-th diagonal element (v_{ii}) of the hat matrix is large, it results in a residual value with a small variance and that means that the predicted value is nearly equal to the observed value. Such cases are important in determining the β's, and v_{ii} is called the *leverage* of the i-th data point.

Least squares is the optimal procedure when the errors are normally distributed but a long ways from best when they are not normal. Fitting techniques are *resistant* if the result does not change greatly when a small fraction of the data are altered; they are *robust of efficiency* when the efficiency is high for conditions that fall short of normally distributed errors and equal variances throughout. In such cases *robust regression* is a nice alternative. The residuals from an estimate $\mathbf{b}$ of the vector $\boldsymbol{\beta}$ of regression coefficients will be denoted by $r_i(\mathbf{b})$ and the median of the absolute residuals by $s(\mathbf{b})$. A new estimate of the parameters is formed by finding a local maximum of $\Sigma\psi(r_i(\mathbf{b})/s(\mathbf{b}))$, where $\psi(z)$ is a function of z, namely, $\psi(z) = 1 + \cos(z/c)$ if $|z| \leq c\pi$ and zero elsewhere (c is a constant chosen by the investigator).

The *sensitivity* of an estimator can be defined as the maximum possible influence of a single observation on (1) the estimated parameter vector or (2) its own fitted value or (3) the set of scaled linear combinations of the parameter estimates. A robust regression technique in which an appropriately chosen set of weights is used to bound the influence function is called *bounded influence regression*.

Diagnostics

A number of diagnostic procedures that reveal inappropriate assumptions or anomalies in the data are available. The most common diagnostic technique is a *residual plot*. It consists of plotting the i-th residual against either the fitted value $\hat{y}_i$ or the value of the j-th carrier x_{ij}.

Since the residuals have different variances, it has been recommended that the residual, r_i, be replaced by the *studentized residual*, $t_i = r_i/\hat{\sigma}(1-v_{ii})^{1/2}$, where v_{ii} is the i-th diagonal element of the hat matrix. In either case the interpretation of the residual plots is the same.

A *rankit plot* is a plot of the ordered residual denoted by $r_{(i)}$, against the expected values of the order statistics from an iid $N(0,1)$ sample of size n, which can be found in tables. Rankit plots are used to judge deviations from normality. If the data are normal, the points in the rankit plot should be approximately linear.

In the linear regression model $\mathbf{Y} = \mathbf{X}\boldsymbol{\beta} + \mathbf{e}$, a solution for $\hat{\boldsymbol{\beta}} = (\mathbf{X'X})^{-1}\mathbf{X'Y}$ requires that the columns of $\mathbf{X}$ be linearly independent (i.e., no column of $\mathbf{X}$ can be expressed as a linear combination of other columns). When such a dependency exists, $(\mathbf{X'X})$ is *singular* (cannot be inverted) and the determinant of $(\mathbf{X'X})$ is zero. If the columns of $\mathbf{X}$ are almost dependent, $(\mathbf{X'X})$ is said to be *ill* (or *badly*) *conditioned* and finding an inverse is quite difficult. As a consequence, some parameters are highly correlated with others and can vary widely in the model with little change in the sum of squares; the model is *overparameterized* or *overspecified* in the sense that a model with fewer parameters will lead to a solution. Additional data gathering is an alternative means to getting a solution for an ill-conditioned problem. Dependencies are hard to spot. Let the model be written as $\mathbf{Y} = \boldsymbol{\beta}_0\mathbf{1} + \mathbf{X}\boldsymbol{\beta} + \boldsymbol{\epsilon}$, where $\mathbf{1}$ is an $n \times 1$ column of 1's and $\mathbf{X} = [\mathbf{C}_1, \mathbf{C}_2, \ldots, \mathbf{C}_p]$, where the columns are *centered* (add to zero) and are *standardized* so that $\mathbf{C'}_j\mathbf{C}_j = 1$. If column dependence exists, the process of constructing $\mathbf{X}$ in that form will lead to a column of zeros. Ill conditioning is indicated by a column consisting of very small numbers. Another procedure is to write $(\mathbf{X'X})$ as a matrix of correlations and to find the eigenvalues of the correlation matrix. Zero (or small) eigenvalues indicate dependencies (or near dependencies). A *multicollinearity* (or *collinearity*) is an approximate linear dependence of the columns of $\mathbf{X}$. The closer the columns of $\mathbf{X}$ are to dependency, the greater the *degree of collinearity*. Collinearities result in least squares estimators, $\hat{\boldsymbol{\beta}}$, of $\boldsymbol{\beta}$ with relatively large variances. The *variance inflation factors*, *VIF*, are the diagonal elements of $(\mathbf{X'X})^{-1}$.

A *biased estimator*, $\tilde{\beta}$, will be superior to the least squares estimator, $\hat{\beta}$, if it has sufficiently small *mean squared error (MSE)*, defined as

$$\text{MSE}(\tilde{\beta}) = E[\tilde{\beta} - \beta)'(\tilde{\beta} - \beta).$$

While the least squares estimator has the smallest variance in the class of unbiased estimators, it may be possible to find a slightly biased estimator with a much smaller variance so that mean squared error (bias^2 + variance) is smaller for the biased estimator than for the unbiased one. An alternative criterion for constructing a biased estimator is the *generalized mean squared error, GMSE*, defined as

$$\text{GMSE}(\tilde{\beta}) = E[(\tilde{\beta} - \beta)' \ \mathbf{B} \ (\tilde{\beta} - \beta)]$$

for some symmetric positive (semi) definite matrix B.

A class of useful biased estimators is provided by the general form

$$\tilde{\beta} = \sum_{j=1}^{p} a_j c_j \mathbf{l}_j$$

where $\{a_j\}$ are constants that depend on the particular estimator; the $\{\mathbf{l_j}\}$, $j = 1, 2, \ldots , p$, are the characteristic vectors of $\mathbf{X'X}$, and $c_j = \mathbf{l}'_j \mathbf{X'Y}$. The least squares estimator is of that form and is obtained by setting $a_j = \lambda_j^{-1}$, where $\lambda_1 \leq \lambda_2 \simeq \ldots \simeq \lambda_p$ are the characteristic roots of $X'X$.

The *principal component estimator* is defined by $a_j = 0, j = 1, 2, \ldots , r$, and $a_j = \lambda_j^{-1}, j = r + 1, \ldots , p$. The r terms with $a_j = 0$ correspond to the collinearities. The name is chosen because the estimator retains only the principal (nonnegligible) elements of $(\mathbf{X'X})$.

If we replace each variable in the regression by a new variable, obtained from the old one by subtracting its mean and dividing by its sum of squares, the $\mathbf{X'X}$ matrix is replaced by the correlation matrix of the independent variables, $\mathbf{Z'Z}$. When the $\mathbf{X'X}$ matrix is ill conditioned, we may add some small positive number k to the diagonal elements of $\mathbf{Z'Z}$ and obtain a solutions, $\tilde{\beta}_{RR}(k) = (\mathbf{Z'Z}) + k\mathbf{I})^{-1} \mathbf{Z'Y}$. That procedure is called *ridge regression*. The matrix $\mathbf{Z}$, which replaces the $\mathbf{X}$ matrix, is said to have been "centered and scaled." In the principal component analysis, each new variable Z_j is replaced by a new variable W_j, which is the j-th principal component of the $\mathbf{Z}$'s.

For a fixed constant $k > 0$, an *(ordinary) ridge regression estimator*, $\tilde{\beta}_{RR}$, is obtained by setting $a_j = (\lambda_j + k)^{-1}, j = 1, 2, \ldots , p$, where k is called the *ridge parameter*. Ridge regression thus reduces the influence of multicolinearities rather than eliminating them, as was the case with a principal component analysis. A *ridge trace* is a plot of the value of each component of $\tilde{\boldsymbol{\beta}}_{RR}$ versus k. The ridge trace will have 1 curve, or trace, per component. A *generalized ridge regression estimator* is of the form $a_j = (\lambda_k + k_j)^{-1}$, where $k_j < 0$ are fixed constants, $j = 1, 2, \ldots, p$.

A *latent root estimator* is a biased estimator that is not in the class of estimators specified above. Let $\mathbf{A} = (\mathbf{Y}_s, \mathbf{X})$ where $\mathbf{Y}_s$ is the vector of responses standardized so that $\mathbf{Y}'_s\mathbf{1} = 0$ and $\mathbf{Y}'_s\mathbf{Y}_s = 1$. Also, let $\alpha \leqslant \alpha_1 \leqslant \ldots \leqslant \alpha_p$ and $\gamma_0, \ldots, \gamma_p$ denote the characteristic roots and characteristic vectors of $A'A$ respectively, and $\gamma'_j = (\gamma_{0j}, \gamma_{1j}, \ldots, \gamma_{pj}) = (\gamma_{0j}, \delta'_j)$. A latent root estimator is defined by $\boldsymbol{\beta}_{LR} = \sum_{j=t}^{p} f_j \delta_j$ where

$$f_j = -\sqrt{\sum (y_1 - \bar{y})^2} \, \gamma_{0j}\alpha_j^{-1} / \sum_{i=t}^{p} \gamma_{0i}^2 \alpha_i^{-1}$$

and t is an integer $0 \leqslant t < p$.

A *shrunken estimator* is of the form $a_j = d\lambda_j^{-1}$, where $0 \leqslant d < 1$. For $p \geqslant 3$, the *James-Stein shrunken estimator* is given by

$$d = \max\{0, \, 1 - f(\text{RSS})/\hat{\boldsymbol{\beta}}'\hat{\boldsymbol{\beta}}\}$$

where $0 < f < 2(p-2)/(\text{RSS} + 2)$ and RSS is the residual sums of squares from a least squares regression. The mean squared error for the James-Stein shrunken estimator is minimized by choosing $f = (p - 2)/(\text{RSS} + 2)$.

Isotonic regression is a name given to regression (or estimation) under order restrictions. As an example, suppose we wish to test a group of means for no trend versus an alternative of a nondecreasing trend. Then $H_0: \mu_1 = \mu_2 \ldots = \mu_n$ and $H_A: \mu_1 \leqslant \mu_2 \leqslant \ldots \leqslant \mu_n$. Suppose also that we have 1 observation from each population. The observation from the i-th population is denoted by y_i. Under H_0 the estimate of $\mu_i = \mu_2 = \ldots \mu_n$ is $\Sigma y_i/n$; under H_A the estimate is $\Sigma(y_i - \mu_i)^2$ subject to $\mu_1 \leqslant \mu_2 \leqslant \ldots \leqslant \mu_n$. That is a quadratic programming problem subject to linear restrictions.

Subset Selection and Restricted Least Squares

A *restricted* or *constrained least squares estimator* is an estimator chosen to minimize

$$(\mathbf{Y} - \mathbf{X}\boldsymbol{\beta})' \, (\mathbf{Y} - \mathbf{X}\boldsymbol{\beta})$$

subject to the consistent linear restrictions $\mathbf{H}'\boldsymbol{\beta} = \mathbf{L}$, where the rank of $\mathbf{H}$ is $s \leq p+1$. A restricted least squares estimator may be biased or unbiased. *Subset selection* is a method of choosing a restricted least squares estimator when $\mathbf{L} = 0$ and $\mathbf{H}'$ is some column permutation of the matrix $(\mathbf{I}_{sxs}, \mathbf{0})$. Typically $\mathbf{H}$ and s are data dependent so that the estimates constructed via subset selection are generally biased. Subset selection may be viewed as a method of constructing an estimator with a small mean squared error. There are 4 basic approaches to subset selection: *all possible regressions, forward selection, backward elimination,* and *stepwise selection.*

The *all possible regressions* method requires the computation of a criterion of every subset with those subsets that "optimize" the criterion being selected for further consideration. One criterion is Mallow's C_p statistic which, for a p term subset model, is defined as

$$C_p = \frac{\mathrm{ESS}_p}{\hat{\sigma}^2} + 2p - n$$

where ESS_p is the error sums of squares for the subset model and $\hat{\sigma}^2$ is the unbiased estimate of σ^2 from the full model. Subsets with C_p small or at least $C_p < p$ are preferred. The C_p statistic is an estimate of the *total error,*

$$\Gamma_s = \sum_{i=1}^{n} \mathrm{MSE}\ (\hat{y}_{i,p})/\sigma^2$$

where $\hat{y}_{i,p}$ denotes the ith fitted value for the subset model.

In the *forward selection* procedure, the carrier with the highest simple correlation with the response variable is chosen first, and then carriers are added, one at a time, until one of a number of possible criteria is met.

The *backward elimination* method is similar to forward selection, except that the full model is chosen first and then carriers are removed one at a time. At each step the carrier to be removed is chosen to be the one with the smallest value of $\hat{\beta}_j^2/\sqrt{\hat{\mathrm{Var}}(\hat{\beta}_j)}$, $j = 1, 2, \ldots, s$, where s denotes the size of the present subset.

The *stepwise algorithm* is a combination of forward selection and backward elimination. At each step 4 alternatives are considered: add a carrier, delete a carrier, exchange 2 carriers, or stop.

Models With Random Carriers

In the standard regression model it is assumed that the independent variables or carriers are fixed, nonrandom variables that are measured without error. A number of alternative models in which that assumption is relaxed have been investigated.

Suppose that the carriers are measured with error so that, instead of observing x_{ij}, we observe $X'_{ij} = x_{ij} + \delta_{ij}$, where δ_{ij}'s are independent random variables with mean zero. Thus, $EX'_{ij} = x_{ij}$, and the linear regression model relating y_i and x_{ij} may be written

$$y_i = \beta_0 + \beta_1 EX'_{i1} + \beta_2 EX'_{i2} + \ldots + \beta_p EX'_{ip} + \epsilon_i,$$

where ϵ_i is assumed independent of X'_{ij}, $i = 1, 2, \ldots, n; j = 1, 2, \ldots, p$. It is called the *error-in-variables model* or the model for *regression with both variables subject to error*. Alternatively, that model may be written as

$$Ey_i = \beta_0 + \beta_1 EX'_{i1} + \beta_2 EX'_{i2} + \ldots + \beta_p EX'_{ip}.$$

Such relationships between expected values of random variables are termed *functional relationships*.

A *structural relationship* is a relationship, not between the expectations of random variables, but between the random variables themselves. For example, if $V, U_1, \ldots, U_p$ denote random variables, then a *structural linear relationship* between V and $U_1, \ldots, U_p$ is

$$V = \beta_0 + \beta_i U_i = \ldots + \beta_p U_p$$

where the β_i's are fixed, unknown parameters. If V is unobservable, but $U_1, \ldots, U_p$ and $y = V + e$, where $E(e) = 0$ and $Var(e) = \sigma^2$, are observable, then

$$y_i = \beta_0 + \beta_1 U_{1i} + \ldots + \beta_p U_{pi} + e_i.$$

That is called a *structural linear model*. If the U_{ij}'s are unobservable, but $X_{ij} = U_{ij} + \delta_{ij}$ is observable, then we may form a *conditional errors-in-variable* model,

$$E(y_i|\{U_{ij}\}) = \beta_0 + \beta_1 E[X_{i1}|U_{i1}] + \ldots + \beta_p E[X_{ip}|U_{ip}].$$

Estimation techniques in structural or functional models are generally more complicated than those associated with least squares and usually require *a priori* information. One notable exception to that is the *controlled variables model*, which is also known as the *Berkson model*. In the Berkson model the carriers are random variables, but their observed values are controlled by design. For example, in the simple linear case, the single carrier may correspond to a dial setting, which is controlled by the experimenter. If the dial setting is x but the true and unknown setting is $X = x + \gamma$, then the observed value has been controlled but the true value is a random variable. The effect of that control is to reduce the underlying functional model

$$y = \beta_0 + \beta_1 (x + \delta) + \epsilon$$

to a standard least squares model

$$y = \beta_0 + \beta_1 x (\beta\delta + \epsilon)$$
$$= \beta_0 + \beta_1 x + \epsilon'$$

Finally, a *components of variance model* is one in which the regression coefficients, $\beta_1, \ldots, \beta_p$, are assumed to be random variables. In those models interest centers on inferences about the variances of $\beta_1, \ldots, \beta_p$.

In regression and discriminant analysis it has been desirable to "check" or "validate" the model. For that purpose the data are split into 2 parts, one of which serves as a *training set* for constructing the model and the other is used to measure the performance of the model. The data must be divided at random.

The *PRESS* routine (predicted residual sum of squares) is used in a regression context to make a choice between several models. For a sample size of n, a given model is fitted n times, leaving out 1 point each time. The missing data point is predicted each time by inserting the regressors into the fitted model. The sum of squares of deviations (observed minus predicted values) is then calculated. That is repeated for each model. The model with the smallest sum of squares is then chosen as the "best" model. Similar schemes leave out m points each time, with $m = n/2$ being a popular choice.

A technique with the same flavor is *cross-validation*, where the response is usually binary (0 or 1), as in logistic regression or discriminant analysis (lived or died, belongs in group 1 or group 2). A prediction rule is devised, which is also binary. Leaving out 1 point at a time, a model is fitted to the remaining $n-1$ points. A function Q_i is constructed, which is zero if the observed value and the predicted value agree and 1 if they do not agree. The average value of Q (over all n points) is the *cross validation error rate*, which estimates the probability of misclassifying a new randomly selected observation.

Similar to that is *jackknifing*, so named because it is a "rough-and-ready tool like a boy scout's trustworthy jackknife." The n data points are divided into g groups with h points per group (h is frequently taken to be 1). We define the *pseudo-values* to be $\tilde{\theta}_i = g\hat{\theta} - (g-1)\hat{\theta}_{-i}$, where $\hat{\theta}_{-i}$ is the estimator of θ with the i-th group deleted, and i ranges from 1 to g. The average of the pseudovalues $\bar{\theta}$ is the *jackknife estimator* of θ. If $\hat{\theta}$ is a biased estimator of θ, $\bar{\theta}$ will be nearly unbiased and the *jackknife variance* is $\Sigma(\tilde{\theta}_i - \bar{\theta})^2/(g(g-1))$. The jackknife is used both for bias reduction and interval estimation (which requires the variance).

Still another method along the same lines is *bootstrapping*, which is proving to be the best of the sample reuse methods. We have a sample, $X_1, X_2, \ldots, X_n$, where each X_i consists of a response y_i with regressors $t_{1i}, t_{2i}, \ldots, t_{pi}$. From the sample, an empirical cdf, $\hat{F}$, is constructed, and we sample *with replacement* from $\hat{F}$. In practice that means that we construct a very large data set by copying the sample values many times. We then sample from the large data set so that the probability of drawing X_i is $1/n$. (In jackknifing we draw samples of size n-1 without replacement). The parameter θ to be estimated can be anything, such as a correlation coefficient, whose distribution and variance are difficult to estimate. From a sample of size n we compute the usual estimate of θ, say θ_1^*, and call it a *bootstrap replicate*. We repeat the procedure to get $\theta_1^*, \theta_2^*, \ldots \theta_N^*$. The *bootstrap estimate* is the average of those values, θ^*, and the *bootstrap variance* is $\Sigma(\theta_i^* - \theta^*)^2/(N-1)$. By ordering the θ_i^*, we can get a *bootstrap distribution* of θ^*. A function of the sample and the parameter θ, which has a distribution independent of θ, is described as *pivotal*, the statistic $(\tilde{\theta} - \theta)/(\text{var}\tilde{\theta})^{1/2}$ being the classic example. Some care must be exercised in using the bootstrap with statistics that are not pivotal.

The most widely used method of estimating the variance of a statistic that is a function of several variables is to expand the statistic into a Taylor series, drop all but the first order terms, and use the rules for linear combinations of variables to estimate the variance. That is known as *propagation of error* or the *delta method*. For $\theta = f(X,Y)$, say, the variance of θ is approximately $(\partial\theta/\partial X)^2 \hat{\sigma}_x^2 + (\partial\theta/\partial Y)^2 \hat{\sigma}_y^2 + 2(\partial\theta/\partial X)(\partial\theta/\partial Y)\hat{\sigma}_{xy}$. The partial derivatives are evaluated at sample estimates of the means of each quantity involved.

A *bioassay* is a measurement of the toxicity or potency of such things as insecticides, vitamins, and hormones when administered to living animals. For vitamins, weight gain is measured, and for hormones, changes in the blood are observed. With insecticides, there is an all-or-nothing response (death) called a *quantal response*. For quantal response data, one objective has been to estimate the LD_{50} (the concentration or dose that kills 50 percent of the insects) or the *median effective dose*, ED_{50}, in case the response is not death. Other percentiles can also be estimated. The experiment is usually performed by giving the first group of insects a toxic material of concentration X_1, the second group, a concentration of X_2, etc., and observing the proportions $p_1, p_2, \ldots, p_m$ in each group that dies. The method used for analyzing the data is *probit analysis*. The model specifies that the expected proportion dying at dose x is $\Phi(\alpha + \beta x)$, where Φ is the cumulative normal distribution. The probit (or probit transformation) of a proportion P is the abscissa Y, which yields an area under the curve of P when a normal curve with mean 5 and standard deviation 1 is used. The mean of 5 prevents Y from being negative. When the probit Y_i is plotted on a linear scale versus the dose X_i, the sigmoidal shape of the cumulative normal is transformed to a straight line and a weighted linear regression can be fitted to the data. The slope of the line is used to estimate the standard deviation.

Another type of estimation of the 50th percentile is the *Bruceton Method (up-and-down method, staircase method)*, used especially for testing how sensitive explosives or other gadgets are to dropping. Starting with a height believed to be near the 50 percent point, we drop an item from the sample. If it explodes, the distance dropped is decreased successively until a unit fails to explode. The distance is then increased(in smaller steps) until another unit explodes. In that way we narrow in on the median or mean (a normal distribution is assumed). Ordnance items that do not explode are tested after each drop to see whether they still function.

The Bruceton method is one form of a more general *stochastic approximation* of the percentiles of a general distribution or a regression parameter. The lower percentiles are much harder to estimate, because the response rate is so much lower. It is not unusual for many thousands of units to be required. An efficient method is the *Robbins-Munro Method*. If we let $F(x)$ be the cdf, we wish to find θ, where $F(\theta) = p$, the desired quantile ($0 < p < 1$). We let X_n be the best guess of θ, and let $y_n = 1$ if there is a response and $y_n = 0$ if there is no response. We then let the next guess be $x_{n+1} = x_n + a_n(p - y_n)$, where a_n is taken to be $1/n$ or a multiple thereof. The *Kiefer-Wolfowitz Procedure* is another method of stochastic approximation.

We shall now discuss the recent but important concept of *generalized linear models*. In the classical regression model, we represent the i-th data point as $Y_i = \beta_0 + \beta_1 X_{i1} + \ldots + \beta_p X_{ip} + \epsilon_i$, where Y_i is a random variable with mean $\mu_i = \beta_0 + \beta_1 X_{i1} + \ldots + \beta_p X_{ip}$, the β's are unknown constants to be estimated, and the X's are known and can be measured without errors. The errors denoted by ϵ_i are assumed to be independent with constant variance σ^2. For hypothesis testing, we added the assumption that the errors were normally distributed. We could as well have broken the problem into 3 components: (1) a *random component* Y_i, which was i.i.d. $N(\mu_i, \sigma^2)$, (2) a systematic *linear predictor* $\eta_i = \beta_0 + \beta_1 X_{i1} + \ldots + \beta_p X_{ip}$, and (3) a *link*, which in that case is $\mu_i = \eta_i$. In the generalization to follow, the linear predictor η_i will be a function of the mean μ_i rather than identical to it. That function will be required to be monotonic and differentiable. We will further allow the random component to come from an exponential family rather than restricting it to a normal family. A family of probability density functions is of the *exponential family* or *exponential class* if $f(x;\theta)$ can be written as $\exp[p(\theta)k(x) + S(x) + q(\theta)]$. The exponential family includes the normal, Poisson, Binomial, Gamma, and Inverse Gaussian. For bounded θ, the statistic $T = \Sigma K(x_i)$ is a complete sufficient statistic for θ. The form for the pdf to be used in the sequel is $f(y;\theta,\phi) = \exp\{[y\theta - b(\theta)]/a(\phi) + c(y,\phi)\}$, where *a, b,* and *c* are specific functions. With ϕ known, that is an exponential family. With this notation, $E(Y) = b'(\theta)$ and $\mathrm{Var}(Y) = b''(\theta)a(\phi)$; ϕ is called the *dispersion parameter*, and $b''(\theta)$ is the *variance function*. The function $a(\phi)$ is often expressed in the form ϕ/w, where w is the *prior weight* and is assumed known. The link relates the linear predictor to the expected value μ_i of a data point y_i. There are several types of link: (1) the logit link: $\eta = \log(\mu/(1-\mu))$, (2) the probit link: $\eta = \Phi^{-1}(\mu)$, (3) the complementary log-log link: $\eta = \log[-\log(1-\mu)]$, and (4) the power family link: $\eta = (\mu^\alpha - 1)/\alpha$. For the principal distributions mentioned above, the canonical links are (1) normal: $\eta = \mu$, (2) Poisson: $\eta = \ln \mu$, (3) binomial: $\eta = \ln(\mu/(1-\mu))$, (4) gamma: $\eta = 1/\mu$, (5) Inverse Gaussian: $\eta = 1/\mu^2$. The measure of *discrepancy* or goodness-of-fit proposed is the log of a ratio of likelihoods called a *deviance*. If we let $\hat{\theta}$ be the estimate of θ under the full model with N parameters and $\tilde{\theta}$ the estimate under an intermediate model with p parameters, the discrepancy is written as $\Sigma w_i [y_i(\hat{\theta}_i - \tilde{\theta}) - b(\tilde{\theta}_i) + b(\hat{\theta}_i)]/\phi = $ deviance$/\phi$. The deviances for the particular cases given above are (1) normal: $\Sigma(y-\mu)^2$, which is simply the residual sum of squares, (2) Poisson: $2\Sigma[y\ln(y/\hat{\mu}) - (y-\hat{\mu})]$, (3) binomial: $2\Sigma[y\ln(y/\hat{\mu}] + (n-y)/(n-\mu)]$, (4) gamma: $2\Sigma[-\ln(y/\hat{\mu}) + (y-\hat{\mu})/\hat{\mu}]$, and (5) Inverse Gaussian: $\Sigma(y-\hat{\mu})^2/(\hat{\mu}^2 y)$. Another measure of discrepancy is the *generalized Pearson chi-square*, $X^2 = \Sigma(y-\hat{\mu})^2/V(\hat{\mu})$.

The analysis of variance in regression is replaced by an *analysis of deviance*, since terms in the model are no longer orthogonal and sums of squares for nonnormal distributions are no longer appropriate. Discrepancies replace sums of squares, and a number of tables for different fits might be required. There is no exact theory for the distributions in the analysis of deviance. In the analysis of residuals, the ordinary residuals are replaced by 3 generalizations of a residual: (1) the *Pearson residual* if $r_p = (y-\mu)/\sqrt{V(\mu)}$, (2) the *Anscombe residual* varies with the distribution; it replaces y by $A(y)$ in such a way that the distribution of $A(Y)$ is as normal as possible, and (3) the *deviance residual* is $r_D = \mathrm{sgn}(y-\mu)\sqrt{d_i}$, where Σd_i is the deviance. As an algorithm for fitting the generalized linear models, maximum likelihood estimates of the β_i can be obtained by iterative weighted least squares, but each y_i is replaced by an adjusted dependent variable Z_i, which is a "linearized form of the link function" applied to the y's, while the weights are functions of the $\hat{\mu}$'s.

We have defined the *likelihood* in simple cases to be the joint density of the observed values—the product of the individual densities of the sample points—and have treated it as a function of the parameters. We have used it chiefly to obtain maximum likelihood estimates of the parameters. We now consider some modifications, particularly as applied to regression problems.

If we have a vector **y** of observations that are realizations of a random vector **Y** with density $f(y;\theta)$, we will consider a transformation of **Y** into 2 new variables (V,W), which does not depend on the parameter θ. From the joint distribution of V and W we can obtain the marginal density of V, $f(v,\theta)$, and the conditional density of W given V, $f(w|v;\theta)$. The likelihoods based on those densities are called respectively the *marginal likelihood* based on V and the *conditional likelihood* based on W given V. We can generalize that notion by transforming **Y** into the sequence of random variables $(X_1,S_1,X_2,S_2, \ldots X_m,S_m)$. The full likelihood can be written as

$$\prod_{j=1}^{m} f(x_j|x_1,x_2, \ldots x_{j-1},s_1,s_2, \ldots s_{j-1};\theta) \prod_{j=1}^{m} f(s_j|x_1, \ldots x_j,s_1, \ldots .s_{j-1};\theta)$$

and the second product is called the *partial likelihood* based on S. The conditional likelihood is thus a special case in which only $X_1 = V$ and $S_1 = W$ are present, and the marginal likelihood is the special case in which X_1 is a known constant, $S_1 = V$, and $X_2 = W$. The marginal and conditional types are ordinary likelihoods, but the partial likelihood is not. The partial likelihood is used in place of the full likelihood when the full likelihood is difficult or impossible to calculate. That is the case with the proportional hazards model, where the large number of nuisance parameters makes the calculation of a full likelihood very complicated. (A *nuisance parameter* or *incidental parameter* is one that is not being estimated and "gets in the way" when other parameters are being estimated.)

In other cases of regression the functional form is not known; hence a likelihood cannot be calculated. We let z_i denote the i-th observation with $E(Z_i) = \mu_i$, and $\text{var}(Z_i) = V(\mu_i)$. If the observations are independent and if the expectations are known functions of the parameters $\beta_1, \ldots \beta_p$ and if the variances are known functions of the means, we can define the *quasi-likelihood* $K(z_i,\mu_i)$ by means of the relationship $\partial K(z_i,\mu_i)/\partial\mu_i = (Z_i - \mu_i)/V(\mu_i)$. A quasi-likelihood has many of the same properties of a log likelihood. If Z is from a one-parameter exponential family, then the quasi-likelihood is the log likelihood. When $V(\mu) = 1$ estimation with quasi-likelihoods is identical to least squares. In other cases the quasi-likelihood can be used for estimation, a fact used in some generalized linear models.

We cannot leave this chapter without speaking about Tukey's EDA methodology for studying "sequences"—(x,y)pairs with equally spaced x-values. We remind the reader that an "exploratory" analysis should precede anything else; perhaps its greatest value is in what it can teach us very quickly. If, after plotting the data, we find that there is some curvature in them, Tukey recommends trying to *straighten* (linearize) them by transforming or *re-expressing* one or both variables through a *ladder of transformations* (. . . . ,

$y^3, y^2, y, \sqrt{y}, \log y, -1/\sqrt{y}, -1/y, -1/y^2, -1/y^3 \ldots$). To decide upon a transformation, we should first plot 3 typical points from the original data and then choose something from the ladder in the direction of the "bulging" side of the 3-point plot: If the plot bulges "upward," we choose a transformation higher on the ladder; if it bulges downward, we use a transformation lower on the ladder.

If the straightened data seem to have a nonzero slope, they are said to be *tilted* and can be "flattened out" or *untilted* by plotting the residuals versus x. We can then study that plot for further clues.

Data smoothing plays an important part in understanding the structure of data plotted chronologically. Tukey's basic smoothing procedure is to use running medians of 3 points (i.e., to substitute the median of y_{i-1}, y_i, y_{i+1} for y_i); he symbolizes that procedure with a "3." When it is repeated one or more times, it is a 3R procedure. Those procedures leave out 1 point on each end. For end-smoothing 1 point, we take the median of the 3 following: (1) the actual end-value, (2) the next-to-the-end smoothed value, and (3) the value obtained by a straight line projection of the last 2 smoothed values to 1 point beyond the original end-value. There may then remain a number of 2-point peaks and valleys that create problems. The sequence of points is then *split* (divided into 2 sections) between the 2 points, and end-smoothing is applied to each part separately. When that is done twice, say, we have a 3RSS smoothing procedure. For the monotone parts of the sequence of batch values, we use a *hanning*, which is 2 repetitions of running means of 2 (a moving average with weights equal to 1/4, 1/2, 1/4). We call that procedure an *H*. The *rough* (or the *residual*) is the actual data value minus the *smooth* (smoothed data value).

For smoothing general 2-dimensional data (x-y pairs), Tukey first partitions or *slices* the x-data into fourths, eights, or sixteenths. The slices occur at *letter values* (HMH, E's, or D's). If a schematic plot is constructed for each slice, they are called *parallel schematic plots*. Within each slice we draw a horizontal line across the slice at a value equal to the y-median for that slice. The resulting step function is called the *broken median*. The result of smoothing the broken median is a *middle trace* or *median trace*. If the hinges of the y-data are plotted as horizontal straight lines across each slice, the 2 resulting step functions are *broken hinges*. The *cross medians* are the points within each slice where the x-median intersects the y-median. The *cross hinges* are the points within each slice where the x-hinges intersect the y-hinges except that we pair the upper-x with the upper-y and lower-x with lower-y when the curve is rising and the upper-x with lower-y and lower-x with upper-y when the curve is falling. It does not matter much what we do when the curve is level. Smoothing the data for the cross-hinges results in 2 *hinge traces*, which are the analogues of hinges in a batch data. The median trace is likewise the analogue of the median. The *E*-traces and *D*-traces may be constructed similarly. A plot of all the traces is a *delineation*. The hinge traces and other traces are usually smoothed by first smoothing the (median-lower hinge) and (upper-hinge median) values and plotting the smoothed values above and below the middle traces. For each slice the adjacent points are plotted, and a convex polygon, which includes all the adjacent points, is constructed. It is called the *adjacent polygon*. The polygon, hinge traces, and middle traces together constitute a *wandering schematic plot*. Such plots are used to view general as well as unusual behavior.

References

This fast-growing field stretches all the way from the simple to the very complex. Most statistical texts treat the subject of simple linear regression quite well. A very elementary treatment emphasing the use of statistical packages is given in Younger, M. S. 1979. *Handbook for Linear Regression*, North Scituate, Mass.: Duxbury. The 2 authoritative references are Draper, N. R., and Smith, H. 1981. *Applied Regression Analysis*, 2nd ed. New York: Wiley and Sons, and Graybill, F. A. 1961. *An Introduction to Linear Statistical Models*. New York: McGraw-Hill. Two books that treat the latest developments are Weisburg, S. 1980. *Applied Linear Regression*. New York: Wiley and Sons, and McCullagh, P., and Nelder, J. A. 1983. *Generalized Linear Models*. New York: Chapman and Hall.

6

The Design of
Experiments and the
Analysis of Variance

Experimental Design has reference to the planning of experiments. Such experiments are studies in which the experimenter deliberately controls certain factors, called *treatments*, that may influence the outcome of the experiment. He then observes or measures the result.

Absolute experiments are used for estimating the effects of the treatments, while *comparative experiments* are used for comparing the effects of 2 or more experiments. *Controlled experiments*, which are the subject of experimental design, are to be distinguished from uncontrolled *observational studies*, in which the investigator cannot control the material (geology, astronomy, etc.) but merely decides which phenomena to observe. The uncontrolled observational study lies in the domain of *survey design*. The *experimental units* are the objects to which the treatments are to be applied. The experimental design consists of 2 basic structures. The *treatment structure* is the set of treatments or treatment combinations that evolves from the formulation of the problem in terms of either a set of hypotheses to be tested (for the comparative experiment) or the types of estimates to be made (for the absolute experiment). The *design structure* is the arrangement of the experimental units into homogeneous groups called *blocks*. The experimental design is specified by describing the method of randomly assigning (*randomizing*) the treatments from the treatment structure to the experimental units in the design structure. Randomization is used to eliminate unforeseen biases and to cancel the correlation between adjacent experimental units. *The design of the experiment* deals with the choice of the treatment structure, the choice of the design structure, including the number of experimental units to be used, and the method of randomly assigning the treatments to the experimental units. The experimental design dictates the *model* to be used in the analysis. In constructing the model, it is assumed that there is no interaction between the components of the design structure and the components of the treatment structure. After the experiment is performed, the *analysis of the experiment* consists of estimating the parameters; testing the hypotheses; constructing confidence intervals; and making inferences, conclusions, or decisions.

During an experiment it frequently happens that several homogeneous experimental units receive identical treatments or treatment combinations. Even though treated alike, the individual experimental units do not respond exactly alike—they exhibit an *intrinsic variability* that cannot be eliminated or controlled. That *inherent variability* is often referred to as *experimental error*, a convenient name for all sources of variation that elude control. A *uniformity trial* is an experiment in which all experimental units are subjected to the same treatment for the sole purpose of evaluating the experimental error. Factors such as pressure, temperature, and time may have an effect on the response of an experimental unit. When they are recognized formally as part of the treatment structure and are allowed for in the analysis, they are called *assignable effects*. Components such as batches, ovens, and litters make up groups of homogeneous experimental units and may also have an effect on the response of the experimental units. They are part of the design structure of the experiment and, though not generally of interest to the experimenter, must be controlled. The remaining uncontrollable or unassignable variation is called *residual variation*. In practice, the effects of lack of uniformity in the physical conduct of the experiment become part of the experimental error or *noise* in the system.

Replication has been called 1 of the 3 basic principles of experimental design. A *replication* of a treatment combination is an independent observation of the treatment combination under conditions as nearly identical to the original as the nature of the experiment will permit. Replication makes possible interval estimates of parameters and tests of hypotheses by providing an estimate of experimental error and increases the sensitivity of the test by reducing the variance of the difference between 2 treatment means. Replication should not be confused with *subsampling* or *repeated measures*, in which there are several responses from the same experimental unit; they do not provide information about the variation between experimental units, which is called *experimental error*. *Randomization* is the process of randomly assigning treatment combinations to experimental units; it calls for making the assignments with equal (or at least known) probabilities. In theory, the best comparisons between treatment combinations are made when the trials are carried out on a set of homogeneous experimental units whose responses differ only because they are assigned different treatments. When the number of trials in an experiment is large, it may be difficult to obtain the necessary set of homogeneous experimental units. It may then be possible and desirable to split the experimental units into smaller groups of more homogeneous experimental units. Such groups are called *blocks* and are characterized by the fact that the "within-block" experimental error is smaller than that over the whole set of experimental units. Such a procedure is called *blocking* and is the basic element of the design structure.

The type of blocking performed specifies the design structure. We now discuss common design structures. When the experimental units are deemed to be homogeneous, there is only 1 block and the treatments are assigned completely at random to the experimental units. That design structure is said to be a *completely randomized design*. The use of blocking imposes some restrictions on the randomization. If each treatment combination is to be assigned to an experimental unit within each block, then the block is called a *complete block*. If the *block size* is less than the number of treatment combinations, then the block is called an *incomplete block*, since not all treatment combinations can be assigned to experimental units in the block. A *randomized complete block* design consists of 2 or more blocks. Each block of the design has as many experimental units as there are treatment combinations (or a multiple of that number). Each treatment combination is randomly assigned to an experimental unit within each block.

The source of variation resulting from blocking is removed with a consequent reduction in the experimental error. In other words, when a blocking scheme is used, the experimental error is measured from within-block comparisons rather than from comparisons among all the experimental units. It is to be noted that for the randomized complete block, each block provides a complete replication of the treatment combinations, but, in general, blocks and replications are not synonymous. The randomized complete block is called a design that blocks in *1 direction* or removes 1 source of variation. The following designs block in 2 or more directions or remove 2 or more sources of variation. In the *Latin square* design (so called because Latin letters are used to denote treatments) the experimental units fall into a two-way layout consisting of "rows" and "columns." Each row is a complete block. In addition, each column is a complete block. The treatments are assigned to the blocks in such a way that each treatment occurs once and only once in each row block and once in each column block. The *Graeco-Latin square* design consists of 3 types of blocks: rows, columns, and tiers (usually denoted by Greek letters). Each block type has the same number of experimental units as treatments, and each treatment occurs once and only once in each row, column, and tier. The *hyper-Graeco-Latin square* consists of more than 3 types of blocks where each treatment occurs once and only once in each type of block. The randomization for the Latin-square-type designs is achieved by selecting 1 of the possible arrangements at random.

There are several mathematical properties associated with the Latin square type arrangements. Latin squares are said to be in *standard form* if the Latin letters in the first row and column are ordered alphabetically or numerically. Two standard squares are *conjugate* if the rows of one are the columns of the other. A square is *self-conjugate* if its arrangement in rows and columns is the same. A permutation of rows, columns, and letters in a Latin square is an *adjugate set*, and if a square is its own adjugate set, it is said to be *self-adjugate*. Two Latin squares are *orthogonal* if, when superimposed, every letter of one square occurs once and only once with every letter of the other.

Before proceeding further, let us discuss the importance of the choice of the treatment structure, which consists of the *factors* (treatments to be studied), neglecting for the moment the method of assigning the treatments to the experimental units. The various *values* of a factor (*amounts* when the factor is quantitative and *classes* or *categories* when the factor is qualitative) are known as *levels* of the factor. A treatment constructed by taking combinations of levels of 2 or more factors is called a *treatment combination*. A group of treatments that contains every combination of 2 or more factors (each at 2 or more levels) is called a *factorial* arrangement or a complete set of treatment combinations. The *dimensions* of a factorial arrangement are indicated by the number of factors and the number of levels for each factor. In a pxq factorial arrangement, there are 2 factors, the first having p levels and the second having q levels. A p^n *factorial* arrangement consists of n factors, each having p levels. The 2^n and 3^n factorial arrangements are widely used. The 2^n factorials are sometimes analyzed with the aid of a *half-normal plot* (defined elsewhere). A $p^n x\ q^s$ *factorial* arrangement contains n factors at p levels and s factors at q levels. A factorial arrangement where not all factors have the same numbers of levels is called a *mixed factorial arrangement*.

The advantage of a factorial arrangement is that the effects of several factors can be investigated simultaneously; a *disadvantage* in their use is that the number of treatment combinations increases rapidly with the number of factors or the number of levels of 1 factor. At the same time it is usually the case that the block sizes are small or should be small. When the block size is smaller than the number of treatment combinations, then some of the comparisons between the treatment combinations will be indistinguishable from the block effects. Such comparisons are said to be *confounded* with blocks or block effects. The theory of *confounding* consists of efficient ways of splitting the treatment combinations into blocks so that important comparisons between treatments are not confounded with blocks while the unimportant comparisons possibly are. The purpose, of course, is to assess the more important comparisons with greater precision. *Defining contrasts* are those treatment comparisons that are confounded with blocks.

A complete replication of a factorial arrangement is sometimes beyond the resources of the investigator, and some reduction in size is necessary. A *fractional replicate* consists of a fraction of the treatment combinations of a factorial arrangement in a completely randomized design. One way to construct a fractional replicate is to construct a confounded design and then use in the experiment only those treatment combinations in 1 of the blocks. When only a fraction of the treatment combinations are used in the design, 2 or more effects (whether main effects or interaction effects) defined on the complete factorial arrangement will be measured by the same comparison of the treatments. When 1 comparison from a fractional replication corresponds to 2 or more comparisons in the complete factorial arrangement, such comparisons are said to be *aliases*; they have more than 1 name associated with them. Good fractional replications are those where important effects (usually main effects and lower order interactions) have aliases with unimportant effects (usually high-order interactions).

There are various types of arrangements for assigning the treatment combinations to incomplete blocks. A *balanced incomplete block design (BIB)* structure is one in which each block contains the same number of experimental units, each treatment is replicated the same number of times, and each pair of treatment combinations occurs together in a block the same number of times as any other pair of treatments. A partially balanced incomplete block requires the concept of *associate* or *associate classes*. The set of treatments that occur in a block λ_i times pairwise with a specified treatment θ is called the set of ith associates of θ. A *partially balanced incomplete* block design (PBIB) is one in which every block contains the same number of experimental units, every treatment is replicated the same number of times (but does not occur more than once in any given block), and one that has the following restrictions on the associate classes: Given 2 treatments which are i^{th} associates, the number of treatments that are j^{th} associates of the first treatment and k^{th} associates of the second treatment is the same no matter which pair of ith associates we start with. Again note that the treatments in the incomplete block designs can consist of the levels of 1 factor or the treatment combinations from a factorial arrangement or a set of treatment combinations from a pxq factorial arrangement with 1 or more control treatments, etc. *Resolvable* incomplete block designs are those for which blocks can be grouped so that each group is a complete replicate.

An important class of partially balanced incomplete block designs is formed when certain blocks of k experimental units are grouped together. Each group of blocks is required to contain all treatment combinations. Those design structures are called *lattices*. If there are k^2 treatment combinations, then a design with r sets of k blocks of size k where each treatment combination occurs only once in each set is called a *square lattice*. If a square lattice has $k + 1$ sets of blocks, the treatments can be arranged such that every pair of treatments occurs together once in a block; such a design is called a *balanced lattice* (and is also a balanced incomplete block design). A selection of 2, 3, or 4 sets of blocks from a balanced lattice is called a *simple* (or *double, triple,* or *quadruple*) *lattice design* respectively. If there are k^3 treatments and 3 sets of blocks (each forming a replication), the design is referred to as a *cubic lattice*. A *rectangular lattice* consists of $k(k + 1)$ treatments in sets of $k + 1$ blocks of size k. An incomplete Latin square or *Youden square* is a lattice consisting of some of the rows or columns of a Latin square. Treatments used in a square lattice may be named as though they were a k^2 factorial arrangement and analyzed accordingly. Thus square lattices are sometimes called *pseudo factorials*. A factorial arrangement with some comparisons confounded with blocks in a Latin square design is called a *quasi-Latin square*. Differences between row blocks and column blocks of the squares are eliminated or accounted for by the analysis, but the requirement that every treatment combination occurs once in each row block and once in each column block does not hold. If the levels of one factor are applied to the row blocks and the levels of a second factor are applied to the column blocks of a Latin square, the resulting quasi-Latin square is called a *plaid square*. A quasi-Latin square in which the main effect of one factor is confounded with row blocks (i.e., the levels of 1 factor are applied to all experimental units in a block) is called a *half-plaid square*. *Magic squares* and *super magic squares* are Latin squares with additional restrictions imposed on the grouping of treatments into smaller rectangles. When the row blocks correspond to time, *cross-over* and *switch-block* (or *reversal*) designs are used. They consist of repeated (or several) Latin square designs with additional restrictions on the assignment of the treatments to cope with situations where the experiment is known to favor 1 treatment over another, i.e., the treatments are arranged so as to be balanced for *carry-over effects* or to estimate *residual effects* of one treatment from 1 time to the next.

Another special incomplete block design is the *split-plot* type (as called in the biological, physical, and agricultural sciences) or *repeated measure* type (as called in the social sciences). Split-plot type designs consist of a treatment structure with at least two factors, each at several levels where the levels of one factor are assigned to the blocks (thus called *whole plots*) and the levels of the other factors are applied to the experimental units within a block (thus called *subplots*). The blocks are the experimental units for the levels of the whole plot factor while the units within a block are the experimental units for the levels of the subplot treatment. Thus the levels of the whole plot treatment are confounded with blocks. Since there are 2 sizes of experimental units, there are 2 types of experimental error, and it is noted in the analysis of variance table by having a whole plot error and a subplot error. The structures can be extended to involve several sizes of experimental units for split-split-plots, etc. For the repeated measure designs, a person is usually a whole plot experimental unit, and that person is measured repeatedly over time, and the time interval is the experimental unit for the levels of the repeated measure factor. If the experimental units are arranged in a rectangle with row blocks and column blocks, then a *strip-trial design* is constructed for a 3-factor experiment by assigning the levels of the first factor to the row blocks, the levels of the second factor to the column blocks, and the levels of the third factor to the experimental units in the intersection of a row block and a column block.

Response surface methodology involves the study of a response that is a function of several factors with quantitative levels. The mathematical function used to describe the mean response as a function of the factors is called the *response surface*. The experimental design usually used for those investigations is a completely randomized design structure with some specific treatment structures. Those treatment structures are often referred to as experimental designs. It is frequently required to find the levels of the factors that *optimize* (maximize or minimize) the response. Factorial arrangements (particularly fractional replicates of the 2^n series) are used to fit a first-degree polynomial via multiple linear regression. Such an arrangement is called a *first order design*. The regression coefficients then dictate the direction to be taken to optimize the response. The size of the step taken in the given direction is a matter of judgment. Having taken the step, the investigator performs a similar experiment, does another fit, and thus "climbs" the response surface (by the method of steepest ascent) in a sequential, "zigzag" fashion. An approach to the "summit" is indicated by slow changes in the independent variable. While fractional replicates of the 3^n series are sometimes used for that purpose, *composite designs*, formed by adding further treatment combinations to the 2^n factorials, enjoy wider usage. We illustrate the process with a 2^3 factorial design, which may be likened to a cube centered at the origin with treatment combinations at the vertices. If points are added at the center of the cube and at points along each axis of the cube, the result is a *central composite design*. Such designs are used when the experiment suggests the optimum is near the center of the cube. If the experiment suggests the optimum is nearer one of the vertices, a *noncentral composite design* is formed by adding treatment combinations near that vertex. Adding points that ensure that the standard error will be the same for all points the same distance from the origin yields a *rotatable design*. For the 2^3 example, a central composite rotatable design would have added points at the center and at (symmetrical) points on the sphere in which the cube is inscribed. The points would then form a *star*. Having performed the experiment and fitted a quadratic in the several variables, the investigator finds the optimum by setting the derivatives—with respect to each variable in turn—to zero and solving the resulting system of linear equations.

Evolutionary Operation or *EVOP* is a continuous experiment carried on within a full-scale plant. The levels of the variables are changed very slowly (in accordance with some design), and feedback is used so that the process moves steadily toward optimum conditions.

Optimal experimental design is concerned with how to allocate the resources (which treatment combinations should be used) in order to obtain a best estimate of β for the linear model $Y = X\beta + \epsilon$. There are at least 4 definitions of "best." (1) A *D-optimal* design is one that is selected to minimize $|(X'X)^{-1}|$, which is equivalent to minimizing the volume of a confidence ellipsoid for the parameter vector β. (2) An *A-optimal* design is one that is selected to minimize the trace of $(X'X)^{-1}$, which is equivalent to minimizing the mean square error of β. (3) A *G-optimal* design is one that is selected to minimize the maximum variance of a linear combination of the β_i, which is equivalent to minimizing the largest characteristic root of $(X'X)^{-1}$. (4) A design is *E-optimal* if the maximal variance among all best linear unbiased estimators of normalized contrasts is minimal. One area that uses optimal designs is that of *linear mixture models*, where the factors are the components of a mixture where the levels are in terms of proportions and the sum of the proportions of each component must add to 1. The resulting sampling area forms a *simplex*, and thus the corresponding design is a *simplex design* or *extreme vertices design*. Again the experimental design involves selecting a particular treatment structure according to the simplex design and then conducting the experiment in a completely randomized design structure. However, any design structure can be used for the simplex treatment structures.

Each of the experimental designs mentioned previously is "analyzed" by means of a statistical technique referred to as the *analysis of variance* (*ANOVA, AOV*). Analysis of variance is a tool for partitioning the total sum of squares of deviations (from the grand mean) into components corresponding to the several *sources of variation* present in the experiment. The source of variation may consist of both treatment effects and design effects (blocks). The analysis is summarized in an *analysis of variance table*, which consists of a line for each source of variation and a line for the "total," since the component sums of squares and the degrees of freedom usually add to the total sums of squares and total degrees of freedom. For each line the table shows the *source of variation*, the component *sum of squares* (*SS*), the *degrees of freedom* (*df*) corresponding to the sum of squares, the *mean square* ($SS \div df$), and either a column for an *F-ratio* (in the fixed effects case), which will be a ratio of 2 of the mean squares, or an *expected mean square* column (in the random effects case), which shows the expected values of the mean squares.

In any case there is a *model* of the observations. The coefficients of the model can be determined by regression methods, but they are seldom of interest in themselves. The sums of squares resulting from each effect are usually calculated directly and used either for hypothesis testing or estimating variances.

We consider now the simplest model in the analysis of variance (the *one-way layout*) in order to define some of its properties. We assume that we have I normal populations with means $\beta_1, \beta_2, \ldots, \beta_I$ and a common variance σ^2 and samples of size $n_1, n_2, \ldots, n_I$ from the respective populations. Let y_{ij} be the jth observation from the ith population. A model of the data is $y_{ij} = \beta_i + \epsilon_{ij}$, where $\epsilon_{ij} \sim N(O,\sigma^2)$. If μ is the weighted average of the β_i, we can define $\alpha_i = \beta_i - \mu$ and the model is then $y_{ij} = \mu + \alpha_i + \epsilon_{ij}$. That model can be applied to 2 very different situations, which we shall call *Model I* and *Model II*. In Model I the α_i is regarded as a fixed constant, and in Model II the α_i is a random variable, which is $N(O,\sigma_\alpha^2)$. An example will perhaps make things clearer. Suppose a machine shop has purchased 6 machines of 1 type and will be using only those 6 to produce a part with dimension y_{ij}. The shop is interested mainly in whether the 6 machines have different means (we will assume they have equal variances). If the means are different, the production manager will want to know which machine is different from the others and he will set about adjusting the means; there is no interest on his part in any other machine in the world. We refer to this Model I as a *fixed effects model*, meaning that the 6 machines are the only ones of interest and that he will not be generalizing beyond the 6. The company that produced those machines may look at the same data in a different light. It may regard the 6 machines as being chosen at random from its conceptually infinite production line. It wishes to know how much variability there is within a machine and how much variability there is between machines. It regards the α_i as random variables and wishes to estimate σ_α^2 and σ_ϵ^2; hence Model II is called the *random effects model or the components of variance model*. Estimation is accomplished by equating expected and observed mean squares and solving for σ_α^2 or σ_ϵ^2. The company is not interested in those particular 6 machines, but in making generalizations about future machines.

The next step up is that for a *two-way-layout*, which would be used for analyzing an IxJ factorial arrangement. The model is $y_{ijk} = \mu + \alpha_i + \beta_j + \gamma_{ij} + \epsilon_{ijk}$, meaning that the kth observation in the ij^{th} cell is made up of an overall mean (μ) plus a row effect (α_i) plus a column effect (β_j) plus an interaction between the ith row and j^{th} column (γ_{ij}) plus a random error term (ϵ_{ijk}). The ij^{th} *cell* consists of the observations that are simultaneously in the ith row and j^{th} column of the table. There is a population associated with each cell that has a *true cell mean* η_{ij} and *true cell variance* σ_{ij}^2. The ith *row mean* A_i is the average of the true cell means in the ith row and similarly for the jth *column mean* B_j. The *general mean* μ is the average of the row or column or true cell means. The *main-effect* of the ith level of factor A (or the ith *row effect*) is $\alpha_i = (A_i - \mu)$ and similarly for the jth *column effect*, β_j. The *interaction* (γ_{ij}) of the ith level of A with the jth level of B is what remains after subtracting the general mean, the ith row effect, and the jth column effect from the ij^{th} cell mean, i.e., $\gamma_{ij} = \eta_{ij} - (A_i - \mu) - (B_j - \mu) - \mu$. If the interactions are zero, the effects are said to be *additive*. The cell means are thus $\eta_{ij} = \mu + \alpha_i + \beta_j + \gamma_{ij}$.

(For an *exploratory* approach to two-way and higher tables, Tukey recommends a *median polish*. The first step is to find the median of each column and subtract it from the values in that column. Using the resulting data, one finds the medians for each row and subtracts them from the values in the corresponding row. The same procedure is then repeated from the first. Of course one can start with rows instead of columns, and 1 cycle instead of 2 may suffice to get comparable medians for rows or columns. One then studies the residuals as an indicator of basic structure. The medians found in the first 2 steps are comparable to row and column effects in a two-way ANOVA.)

Again the cell means may be fixed so that attention is given only to those rows and those columns, and the experimenter does not wish to generalize. Then the design has *fixed effects*. If rows have been chosen at random from an infinite population, if the same is true of the columns, and if the intent is to generalize, we have a *random effects* model. If rows are fixed and columns are random (or vice versa), we have a *mixed effects model*.

A complete two-way (or higher-way) layout is called a *crossed design*; in such a design every level of factor A occurs with every level of factor B. The design is said to be a *nested design* (with factor B nested in A) if no level of factor B occurs with more than 1 level of factor A. Nested designs are also called *hierarchical designs* (not *hierarchal*). A layout involving several factors that is neither (fully) crossed nor nested is said to be *partly crossed* or *partially hierarchical*.

A special case of an unbalanced layout is one in which the cell numbers in any 2 rows (or columns) are proportional. Such a layout is said to have *proportionate numbers in the subclasses* or to have *proportional frequencies*. Otherwise the layout is said to have *unequal (or disproportionate) subclass frequencies*. Two methods of analysis of the latter case are the *method of fitting constants*, used when the interaction term is assumed to be zero, and the *method of weighted squares of means*, used when interaction is present.

The usual first phase of a Model I analysis of variance is a comparison of the set of population means. If it is decided that there are differences among the set, the second phase is to find out which means differ from each other; that process is referred to as making *multiple comparisons*. A *contrast* is a linear combination $\Sigma c_i \beta_i$ of the parameters (usually means) subject to the restriction that $\Sigma c_i = 0$. The *S-method* (originated by Scheffe) and the *T-method* (originated by Tukey) are methods for judging simultaneously all contrasts. The special case of making pairwise comparisons of means may be handled by the above or by one of the following methods: *Multiple t-tests*, (Fisher's) *Least Significant Difference (LSD)* test, (Student-Newman-Keul's) *Multiple Range* tests, (Duncan's) *Modified Multiple Range* tests, or (Waller-Duncan) *Bayesian* tests.

In pairwise comparisons of means, the contrasts are $\mu_i - \mu_j$ with estimated variance $\hat{\sigma}^2 = s^2(1/n_i + 1/n_j)$, where s^2 is the mean square error in the analysis of variance and n_i is the number of observations associated with the i-th mean $\bar{x}_i$. In the Scheffe method a pair of means differs significantly if the absolute difference exceeds $(F_{1-\alpha,1,k}.\hat{\sigma}^2)^{1/2}$, where k is the d.f. of s^2. In the Tukey Method means differ significantly if the absolute difference exceeds $q_{1-\alpha,k}\hat{\sigma}/\sqrt{n}$, where $q_{1-\alpha,2,k}$ is the $(1 - \alpha)100$th percentile of the studentized range. That method requires that the sample sizes be equal, i.e., $n = n_i = n_j$. The method of multiple t-tests is based on the Bonferroni inequalities and requires that absolute differences in means exceed $t_{1-\alpha/m,k}\hat{\sigma}$, where m is the total number of t-tests to be performed. Fisher's LSD tests are not made unless the F-test in the analysis of variance is significant. Then a pair of means differs significantly if the absolute difference exceeds $t_{1-\alpha/2,k}\hat{\sigma}$. For Duncan's multiple range test, the p means are first ranked and $(p - 1)$ critical values $r_{\alpha,j,k}$ found from tables of the least significant studentized range. The subscript j means that 2 means with $j - 2$ means between them are being compared. If the absolute difference of that pair exceeds $r_j(s^2/n)^{1/2}$, the means are declared significantly different.

The Student-Newman-Keuls procedure likewise requires calculation of $(p - 1)$ critical values from tables of the studentized range, which are multiplied by $\hat{\sigma}$. The Duncan-Waller Bayes Test relies on critical values, which are $\hat{\sigma}$ times a tabled value of the minimum average risk. For pairwise comparisons, Duncan's multiple range, the Duncan-Waller test, and the LSD test (with a preliminary F-test at the .05 level) have been judged superior to the others. For other contrasts with possibly unequal sample sizes, Scheffe's method is the only one available.

References

An old classic is Cochran, W. G., and Cox, G. 1957. *Experimental Designs*. New York: Wiley and Sons. A book with much more philosophy is Cox, D. R. 1958. *Planning of Experiments*. New York: Wiley and Sons. A very thorough treatment is found in Winer, B. J. 1962. *Statistical Principles in Experimental Designs*. New York: McGraw-Hill.

7

Reliability and Survival Analysis

In the production of electrical and mechanical devices, which are required to be very dependable or which wear out, companies spend millions evaluating the reliability of their products. The evaluation includes risks, liabilities, costs of warranties, guarantees, replacements, design changes, etc. All of that involves the study of the *lifetime* of the device in question. In epidemiology or medical experiments (animal and human), lifetime is again a quantity of major interest, and the same methods are used. We need not be concerned solely with the death of the animal or human or with the complete wear-out of the device, but we can apply the same methodology to the time to tumor, time to disease, or *time to occurrence* of any event.

The *operation* of a "unit" takes in all phases of its existence, including transportation, maintenance, servicing, and repair. The *reliability of a unit* is the probability that it will perform its intended purpose adequately for a given length of time under specified conditions. The reliability is a function of time. An urgent problem in our technology has been to increase the reliability of the units, but a great many electrical units now have astounding reliability. The prediction and improvement of reliability of mechanical parts is still a major problem. Basic to understanding reliability is the definition of a *failure*, which is the loss of the properties of the unit so that its functioning is impeded or stopped. Failures are said to be *independent* if they occur in a random fashion with no relationship to other failures; they are *dependent* or *secondary* if they are the result of or the by-product of an independent failure. *Catastrophic failures* are those that occur suddenly and without warning and result in complete mission failure. *Initial failures* occur shortly after the unit is put into operation. *Wear-out failures* result from fatigue, depletion, etc.

The study of survival time is complicated by incomplete data. At any point in time some units will not have failed, and we know only that their lifetime is greater than some number L. Those data are *censored on the right*, and the unfailed units have been called *run-outs, survivors, removals,* or *suspended units*. If we know only that the lifetime was less than L, the observation is *censored on the left*. If all the units are put on test at the same time and the experiment is stopped before the units have all failed, the data are said to be *singly censored*. If that experiment is stopped at a predetermined time, the data are *singly time censored* or *Type I censored*; and if it is terminated when a predetermined number of units have failed, the data are *singly failure censored* or *Type II censored*. For time-censored data the number of failures is a random variable, and for failure-censored data the time to the fixed number of failures is a random variable. In a Type II censoring situation the experiment stops when r failures are observed, leaving $n - r$ unfailed units; if we continue testing n_1 of those units until r additional units have failed and then continue testing n_2 of the remainder, etc., we have *progressive Type II censoring*. If the units are not all put on test at the same time but are censored in one of the above ways, we say that the data are *multiply* (or *hyper,* or *progressively* or *arbitrarily*) *censored*.

If a unit begins to function at time $t = 0$ and fails or dies at time $t = T$, we call the random variable T the *lifetime* or *survival time* or *time-to-failure* of the unit. The distribution of T will be denoted by $F(t)$ and the pdf by $f(t)$. The function $R(t) = 1 - F(t)$ is called the *reliability function* or *survivorship function* or *survivor function* and gives the probability that an individual will survive until time t (or beyond). Related to $R(t)$ is the *hazard function* $h(t) = f(t)/R(t)$, also called the *instantaneous failure rate* or *force of mortality*. Knowledge of any 1 of the 4 quantities $f(t)$, $F(t)$, $R(t)$, and $h(t)$ is sufficient to know all 4. The *hazard function* is the instantaneous failure rate at time t given that the individual survives until time t, and it is of major interest in this field. If $h(t)$ is increasing (decreasing), with time, the distribution is said to be an *increasing (decreasing) failure rate distribution*. In probability terms, $h(t)\Delta t$ is the approximate probability that a unit will die (fail) in the interval $(t, t + \Delta t)$ given that it was alive (or operating) at time t. The *cumulative hazard function* is $H(t) = \int_0^t h(x)dx$ so that $R(t) = \exp(-H(t))$. The hazard function is nonnegative, but $\int_0^\infty h(t)dt = \infty$.

The expected value of T, $\int_0^\infty yf(y)dy = \int_0^\infty R(y)dy$ is the *average* or *expected life* of the units or the *mean-time-to-failure*. The plot of $R(t)$ versus t is called the *survival curve* or the *survival characteristic*. If we let $n(t)$ denote the number of units surviving at time t and N the total number of units tested, the function $R_N(t) = n(t)/N$ is called the *empirical reliability function* and approximates $R(t)$ for large N. If the unit is renewed through repair, the expected value of T is also called the *mean time between failures*.

An important part of reliability theory deals with the choice of the lifetime distribution, the estimation of the parameters of the distribution, tests of hypotheses about the parameters and the construction of confidence intervals for them. The foremost models for survival time are the exponential, Weibull, gamma, and lognormal distributions. Even vague information about the shape of the hazard function can be quite helpful in choosing a model. After it has been chosen, a goodness-of-fit test is frequently used to judge its adequacy. If adequate, the model can then be used to predict reliability and make statements about the reliability.

An area in which there is considerable interest is the identification of factors that may be related to or predictive of survival time, called *risk factors, prognostic variables, explanatory variables,* or *covariates*. There are several regression approaches used for that purpose. If we let t_i be the survival time of the ith individual and $x_{1i}, x_{2i}, \ldots , x_{pi}$ be the values taken on by p explanatory variables for the ith individual, the *proportional hazards model* or *Cox's model* gives the hazard for the ith individual as $h_i(t) = h_0(t)\exp(\Sigma\beta_j x_{ji})$, so that the hazard for any individual is proportional to a baseline hazard function, $h_0(t)$, and the risk factors x_{ji} modify the hazard. The object is to estimate the parameters β_i and test hypotheses about them. When the β_i are significantly different from zero, they indicate that the risk factor is important. That process, sometimes termed *Cox-regression*, is accomplished by maximum likelihood, but a *partial likelihood* (based on conditional probability of failure) is used instead of the usual likelihood function.

For a binary response (does or does not have the disease), the probability of "success" for the ith individual has been modeled as $P_i = \exp(\Sigma\beta_j x_{ij})/(1 + \exp(\Sigma\beta_j x_{ij}))$, where x_{ij} is an explanatory variable or risk factor. That model is the *linear logistic model*, and *logistic regression* has as its object the estimation of the β_i.

It is sometimes useful to portray survival data graphically. The empirical survivor function is useful in that way but needs to be modified when the data are censored. A widely used nonparametric estimate of the survivor function that takes censoring into account is the *product-limit estimate* or *Kaplan-Meier estimate*. If the deaths occur at distinct times $t_1, \ldots , t_k$ with d_j deaths at time t_j, the product limit estimate is $\hat{R}(t) = \Pi(n_j - d_j)/n_j$ where n_j is the number of subjects at risk at time t_j. The product is taken over all j with $t_j < t$.

The *life table* is yet another method of portraying the survival experience of a group of people, referred to as a *cohort*. For each of the successive intervals of time, the table shows (a) number of deaths, (b) number of withdrawals, (c) number at risk, (d) the probability that an individual dies in the interval given that he has survived up to the beginning of the interval, and (e) the probability that he survives beyond the interval.

There is a large body of literature on the subject of plotting the empirical survivorship function, survival density, hazard function, cumulative hazard, etc. *Probability paper* is constructed so that a plot of the ordered data versus the empirical cumulative distribution will approximate a straight line if the data are from the distribution postulated. It is used as a means of choosing a distribution function that fits the data (plots as a straight line). Such a plot is called a *probability plot*. Actually the *plotting position* Y_i should be either the expected value $100(i/(n + 1))$ or $100(i - 1/2)/n$, since it will be impossible (and meaningless) to plot the last point otherwise. If it does fit, the plot is used to estimate the parameters. For the normal the 50th percentile estimates the mean, and the difference between the 50th and 16th percentiles estimates the standard deviation. A *half-normal plot* is sometimes used to interpret the contrasts in a 2-level factorial design or a fractional replicate. The contrasts are ordered and plotted on probability paper for which the lower half of the probability scale is deleted and each probability P on the upper half is replaced by $2P-100$. The 63rd percentile from an exponential plot estimates the mean. For a Weibull or Extreme-value distribution, the 63rd percentile is an estimate of the scale parameter. *Hazard plots* resemble probability plots and are interpreted the same way, but the ordered observations are plotted against the cumulative hazard function.

Given 2 sets of survival data, a major problem is to test the 2 survival functions for differences (males versus females, for example). Nonparametric tests used for that purpose include *Gehan's Generalized Wilcoxon Test*, the *Cox-Mantel Test*, the *logrank test*, *Peto and Peto's Generalized Wilcoxon Test*, *Cox's F-test*, the *Mantel-Haenszel Chi-square test*, and the *Kruskal-Wallis Test*.

If a failed unit is replaced immediately, we say that it is *renewed*. We let $t_1, t_2, \ldots$ denote the instances of failure, and $T_1 = t_1, T_2 = t_2 - t_1, \ldots$ be the times between failure or the *interarrival times*. We then say that a *renewal process* is a sequence of independent and identically distributed nonnegative random variables, $T_1, T_2, \ldots$, which are the interarrival times. We let $F(t)$ and $f(t)$ denote respectively the cdf and pdf of the T_i and $N(t)$ the number of renewals in the interval $(0,t)$ and let $G(t) = E(N(t))$ be the mean or expected number of failures up to time t. $G(t)$ is called the *renewal function*, and its derivative, $g(t)$, is the *renewal density*. The probability that $N(t) \geq n$

is the n-fold convolution $F^n(t)$ of $F(t)$. A basic result in renewal theory is that $G(t)$ satisfies the *renewal equation* $G(t) = F(t) + \int_0^t G(t - s)f(s)ds$. The renewal density is sometimes confused with the *hazard function* $h(t)$, and both have been called the *intensity function*, but $g(t)$ is approximately the *unconditional* probability of failure in a sufficiently small unit of time, while $h(t)$ is approximately equal to the *conditional* probability of failure in the same interval given that there have been no failures before the instant t.

The principal results in renewal theory deal with the fact that $N(t)$ is asymptotically normal so that with the passage of time the process tends to become stationary and lose its dependence on time. *Blackwell's Theorem* states that for continuous $F(t)$ and arbitrary α, the limit of $G(t + \alpha) - G(t)$ as $t \to \infty$ is T_0/α, where T_0 is the mean lifetime of a unit. *Smith's Theorem* or the *Key Renewal Theorem* states that if $Q(x)$ is nonincreasing, the limit of $\int_0^t Q(t-s)dG(s) = T_0^{-1}\int_0^\infty Q(s)ds$, where $Q(x)$ is the distribution of lifetimes. It is so designated because so many results can be proved from it.

A process that requires a nonnegligible amount of time to replace a failed unit is called a renewal process with *finite renewal time* (or an *alternating renewal process*). For such a system the *availability* is the probability that the unit will be functioning satisfactorily at time t given that it was operative at time $t = 0$. For systems that are not time-dependent, that probability is usually thought of as a limiting value of "up time" divided by "total time." The complement of availability is *unavailability*. *Intrinsic availability* is availability when only active repair time is counted as "downtime." *Equipment availability* is the probability that a stated percent of the equipment meets those criteria, i.e., the equipment provides adequate performance for a period of time T (with no downtime interval exceeding the maintenance time constraint). *Mission availability* is defined by substituting *mission* for *equipment*. The *repairability* is the probability that a failed system will be in operation within a given length of active repair time. The *operational readiness* is the probabilty that at time t the system will be operating or could be operational if needed.

For those units whose ability to function is maintained by renewal operations, *maintainability* is the probability that a device will be restored to operational effectiveness within a given period of time provided that the maintenance action is performed in accordance with prescribed procedures. Maintenance has been classified as *emergency maintenance* if required as the result of failure, *preventive* (or *scheduled maintenance*) if periodic and in accordance with specific instructions and scheduling, and *precautionary maintenance* if *not* in accordance with specific instructions or scheduling.

A *system* is made up of *components*. A major problem in reliability is to obtain the reliability of a system given the reliability of the components.

If the components of a system fail independently of one another and if failure of any component causes failure of the system, the components are said to be *connected in series* (in a reliability sense); they are said to be *connected in parallel* if the system fails only when all components in the system fail. *Redundancy* is the existence of more than 1 means for accomplishing a given task, where all means must fail before the system fails. In *parallel redundancy* the components are connected in parallel and operate simultaneously, while in *standby redundancy* some components are on-line and others are standing by idly (off-line), waiting to be switched on when the on-line components fail. The standby components cannot fail until they are switched on. In some cases the system includes *partially energized standby units*, which can fail while off-line, but the probability of doing so is less than that for the on-line units. *System dependability* is the probability that a system will be available between any 2 scheduled maintenances. *System effectiveness* is the probability that the system can meet an operational demand within a given time period. The *maintenance-ratio* is the number of maintenance man-hours of down time required to support each hour of operation. The *mean recurrence time* or *mean time to repair* is the average length of time to return from a failed state to an acceptable state. In that connection the *useful life* of an item is the total operating time between debugging and wear-out, while the *shelf life* is the length of time an item can be stored under specified conditions and still meet specifications or operational requirements. If a device has survived up to time *t*, the mean time to failure is called the *mean residual life*.

Replaceability is the ease with which a part of component can be replaced without requiring extensive disassembly of adjacent hardware. *Serviceability* consists of those properties of an equipment design that make it easy to service and repair while in operation. A *warranty period* is the time, associated with a given probability (*or assurance level*) before which *no* failures will occur in a lot of specified size to be manufactured in the future. More specifically, if a 100λ percent prediction interval is placed on the first failure in a future sample, the lower limit of the interval is called a 100λ percent *warranty period, safe warranty life,* or *assurance limit.* A *guarantee time* is a time before which an individual cannot die. The scale parameter in the Weibull distribution of lifetimes is called the *characteristic life*. The *remaining life* at age *y* is the conditional mean life (given *y*) minus *y*.

Life tests are conducted for purposes of *demonstrating reliability*. If success or failure is measured on each test and if the number of the tests *n* is fixed in advance and are all carried to completion, we speak of *binomial sampling*.

If the units are tested sequentially until a fixed number of units have failed, *n* becomes a random variable and we refer to the process as *Pascal sampling*. If the test time is fixed, failures are replaced as they occur; the number of failures then has a Poisson distribution and the process is called *Poisson sampling*.

High reliability of a device means longer times before failure; hence the measurement of reliability for such devices becomes more difficult (in terms of time, at least). Since measurement of reliability is essential, it is sometimes necessary to obtain data more rapidly than usual by means of *accelerated life tests* in which the device is operated at stress levels higher than normal and the performance under normal stress levels predicted.

Bayesian estimation plays a prominent part in reliability estimation. In the Estimation chapter we discussed prior and posterior densities and Bayes estimators without being specific. Given a prior density $g(\theta)$ and the sampling density $f(x|\theta)$ of the observations given a particular value of θ, we seek the posterior density $g(\theta|x)$ of θ given the sample. *Bayes Theorem* states that if $B_1, \ldots, B_n$ is a collection of mutually disjoint and exhaustive events (the union of the B_i is the sample space) and if all the B_i have positive probability, then for any event A with positive probability, $P(B_k|A) = P(A|B_k)P(B_k)/ \Sigma P(A|B_j)P(B_j)$. The theorem applied to our problem gives $g(\theta|x) = f(x|\theta)g(\theta)/ f(x)$, where $f(x)$ is the marginal pdf of X obtained from $\int f(x|\theta)g(\theta)d\theta$ if X is continuous and $\Sigma f(x|\theta)g(\theta)$ if X is discrete.

An important branch of reliability is that of *risk assessment*. When safety becomes an important factor (nuclear reactors, space flights, manufacture of toxic chemicals), it is necessary to study the *risk* (probability of loss or injury to persons or property) involved. A study comparing the risks of a technology or process with the benefits is called a *risk-benefit analysis*. If the costs (in money, health, lost workdays, etc.) are included, the study is a *cost-risk-benefit* analysis or simply a *cost-benefit* analysis. A study of risk usually proceeds in 3 phases: (1) identifying the hazards, such as an explosion or toxic release, (2) examining the accident sequences that led to the hazard, (3) assessing the degree of hazard (amount of toxin released) and consequences (health effects, property damage).

A *fault tree analysis* starts with the undesirable accident/disaster of interest, labeling that as the *top event*. Through deductive logic, the events that could lead to the top event are listed on a diagram as a layer of rectangles below the top event. If several are required to occur simultaneously before the top event occurs, they are connected to the top event with an *AND-gate*. If the occurrence of any one of several events would cause the top event, they are connected to it with an *OR-gate*. Below that layer of secondary events, we place another layer of events that led to the secondary events and are connected to them by *gate symbols*. (In addition to the AND and OR gates there is an *inhibit gate*, which produces the output event provided that the input event and a conditional event both occur; the *priority AND gate*, which produces the output event if the input events occur in a prescribed order; the *exclusive*

OR gate, which produces the output event if one but not both input events occur; and the *m-out-of-n* or *voting gate*, which results in the output event if *m* out of *n* events occur.) That process is continued until the *basic events*, denoted by circles, occur. The basic events are the limit of resolution of the fault tree: the events that cannot be broken down further or those for which no failure data exist. The *fault tree* is thus a Boolean logic diagram with events of the success-failure type (the valve either opens or fails to open). A *cut set* is a collection of basic events such that if all the events in the cut set occur, the top event is bound to occur. A *path set* is a collection of basic events with the property that if none of them occurs the top event will surely occur. A *minimal cut set* is a collection such that if any event is removed, it is no longer a cut set. One object of the analysis is to identify the minimal cut sets. Each event in a fault tree has a probability of failure associated with it. The object is to find the probability of the top event. That probability will be used to measure the acceptability of the risk involved, usually as an assurance to the public. For new equipment, such as a new reactor, fault tree analyses of existing reactors will indicate events with high risk, and they can be used for designing better or redundant components that will lower the risk.

The fault tree is a *backward analysis* because we work from the top event backward to the basic events. A *forward analysis* begins with a failure event, such as a valve failing to open. All the possible consequences (scenarios) of the failure are then examined and listed above that *initiating event*. Inductive rather than deductive logic is used. The sequence of events resulting from the first event is then diagrammed, and in that fashion an *event tree* is constructed. An event tree analysis fails in case of parallel sequences and is not used as much as fault tree analysis. Forward analysis is also used with a *failure modes and effects analysis (FMEA)*. For each component of the system, a list is made of all the possible ways it can fail and the effect on the system if it does fail. That analysis is more detailed than a fault tree, since many failures will not lead to the top event. A *preliminary hazards analysis (PHA)* lists possible hazards of a system and ranks them as negligible, marginal, critical, and catastrophic. In may include the *accident sequences* that led to certain hazards. A list or diagram is then made of possible corrective actions and consequences of critical and catastrophic failures. A *critical analysis* sometimes follows an FMEA. The component failures are ranked, and on the basis of the ranking, some components are given intensive study or special attention.

A topic of interest in aerospace tracking, underwater sonar, and nuclear safeguards is the *Kalman Filter*, which is like regression in some ways but unlike it in others. Let $Y_1, Y_2, \ldots, Y_t$ be the observed values of a variable at times $1, 2, \ldots, t$. Y_t depends on the state of nature θ_t at time t. The objective is to make inferences about θ_t. The state of nature in that problem is not constant, as is the case with regression, but varies with time. There is a linear relationship $Y_t = F_t\theta_t + \xi_t$, called the *observation equation*, between Y_t and θ_t, where $\xi_t \sim N(0,V_t)$. The dynamic feature (change with time) is described by the *system equation* $\theta_t = G_t\theta_{t-1} + w_t$, where $w_t \sim N(0,W_t)$. V_t and W_t are assumed known, independent of each other, and are called the *observation error* and the *system equation error* respectively. The *Kalman Filter* is the name given to the recursive estimation procedure that uses Bayes theorem:

Pr(state of nature|data) is proportional to Pr(data|state of nature).

References

For a first book I recommend Nelson, W. 1982. *Applied Life Data Analysis*. New York: Wiley. It gives detailed steps to the analysis of most reliability problems and has an excellent set of references. A classic on reliability theory is Mann, N. R., Schafer, R. E., and Singpurwalla, N. D. 1974. *Methods for Statistical Analysis of Reliability and Life Data*. New York: Wiley. A newer, broader treatment is Lawless, J. F. 1982. *Statistical Models and Methods for Lifetime Data*. New York: Wiley. In the area of Bayesian reliability I recommend Martz, H. F., and Waller, R. A. 1982. *Bayesian Reliability Analysis*. New York: Wiley. In the area of risk analysis, an important book is Henley, E. J., and Kumamoto, H. 1981. *Reliability Engineering and Risk Assessment*. Englewood Cliffs, N.J.: Prentice Hall.

8

Order Statistics

By a *random sample*, is meant either (1) a set of n independently and identically distributed random variables, or (2) a sample of n items from a finite population of size N in such a way that the selection of each of the possible $\binom{N}{n}$ samples of size n is equally likely. Let $X_1, X_2, \ldots, X_n$ be a random sample. The *order statistics* of that sample are obtained by rearranging the sample values according to increasing magnitude, so that $X_{(1)} \leq X_{(2)} \leq \ldots \leq X_{(n)}$ with $X_{(k)}$ being called the *k-th order statistic*. Order statistics are widely used, particularly in the following areas: (1) best linear unbiased estimation, (2) estimation where part of the data is missing, (3) distribution free tolerance interval estimation, (4) multiple comparisons, (5) short-cut procedures, (6) nonparametric tests of hypothesis, (7) prediction of extreme events, such as floods and droughts, and (8) detection of outliers.

Order statistics are frequently used to estimate quantiles. The *p-th* (population) *quantile* (or *fractile* or *100p-th percentile*) of the distribution of the random variable X is denoted by ξ_p and is the smallest number ξ for which $F(\xi) \geq p$, where $0 < p < 1$ and $F(x)$ is the cumulative distribution of X; when X is a continuous random variable, the *p-th* quantile is the smallest number ξ satisfying $F(\xi) = p$. The *p-th sample quantile* is $X_{(np)}$ if (np) is an integer and $X_{([np]+1)}$ otherwise, where $[np]$ is the greatest integer less than or equal to (np). Alternatively, the p-th sample quantile is the observation Q_P for which the fraction of X_i's less than $Q_p \leq p$ and the fraction of the X_i's exceeding $Q_p \leq 1\text{-}p$. The 50th percentile, $\xi_{0.5}$, is called the *population median*. The *sample median* is the middle order statistic if n is odd and the average of the middle 2 order statistics if n is even so that it is not quite the sample 0.5 quantile. The *lower quartile* is the 25th percentile $\xi_{0.25}$, and the *upper quartile* is the 75th percentile $\xi_{0.75}$. Thus the median, lower quartile and upper quartile divide the total frequency into 4 equal parts. The *interquartile range* or *interquartile distance* is the difference between the quartiles, $\xi_{0.75} - \xi_{0.25}$. *Deciles* and *quintiles* and *percentiles* divide the total frequency into 10, 20, and 100 equal parts respectively. *Sample quartiles, deciles,* and *quintiles* are the corresponding quantiles estimated from the sample. Quantiles associated with the conventional levels of significance are sometimes called *percentage points* of the distribution.

The largest value or *maximum* is $X_{(n)}$, and the smallest value or *minimum* is $X_{(1)}$. $X_{(n)}$ and $X_{(1)}$ are called the *extremes*, $X_{(n-m+1)}$ and $X_{(m)}$ are the *m-th extremes*, and the largest in absolute value of $X_{(1)}$ and $X_{(n)}$ is the *absolute extreme*. The differences between successive order statistics, e.g. $X_{(i+1)} - X_{(i)}$, is referred to as a *gap* or *spacing*. If we let $\overline{X} = \Sigma X_i/n$ and $S = (\Sigma(X_i - \overline{X})^2/(n-1))^{1/2}$ denote respectively the mean and standard deviation of the sample, then $X_{(i)} - \overline{X}$ is the *i-th deviate* (from the mean), and the largest of $\overline{X} - X_{(1)}$ and $X_{(n)} - \overline{X}$ is the *extreme deviate*. $(X_{(i)} - \overline{X})/S$ is the *i*-th *studentized deviate*, and the extreme deviate divided by S is the *extreme studentized deviate*, the basic statistic used for detecting 1 outlier. The quantity $X_{(1)} + X_{(n)}$ is called the *midsum*, and $(X_{(1)} + X_{(n)})/2$ is the *midrange*. The *range* is $X_{(n)} - X_{(1)}$, while the *i*-th *quasi-range* or *pseudo-range*, denoted by $X_{(n-i+1)} - X_{(i)}$, is the range after deleting $i-1$ observations from each end of the sample. The *geometric range* is $[|X_{(n)}X_{(1)}|]^{1/2}$. The *studentized range* $(X_{(n)} - X_{(1)})/S$ is quite useful in making multiple comparisons. The *j*-th *thickened range* is the sum of the range and quasi-ranges (from 2 to *j*). The quasi-ranges are useful estimators of the standard deviation. The "studentization" referred to is *internal* in the sense of using the sample at hand. *External studentization* is accomplished by using an independent estimate of the standard deviation derived from a separate sample. *Dixon's r-statistics* are ratios of differences of order statistics that are used in the detection of *outliers* (observations not thought to be part of the population). The numerator of Dixon's *r*-statistic is the distance between the suspected observation and its nearest neighbor, while the denominator is the range (or the range with 1 or 2 observations omitted). The studentized range and the largest gap are also used in detecting outliers. The *W-test* is a test of normality; it is the ratio of the square of a linear estimate of the standard deviation (using the order statistics) to a quadratic estimate of the variance (the sum of squares).

A *robust* estimator is one whose properties are insensitive to small deviations from the assumptions upon which the properties depend. Such estimators are particularly useful in *accommodating* (rather than rejecting) outliers. The median is an example. Many such estimators are based on order statistics. A sample is said to be *trimmed* (symmetrically) if the k most extreme observations are omitted from each end of the ordered sample, where k is a predetermined number. A sample is *Winsorized* if the k most extreme observations on each end of the ordered sample are replaced by the nearest retained observation. The α-*trimmed mean*, $m(\alpha)$, is found by discarding a fraction α ($0 \leqslant \alpha \leqslant 1/2$) of the observations from each end of the sample and averaging the remainder (for noninteger values of $n\alpha$ that requires fractional weights for the first and last observation). The average of the discarded values is denoted by $m^c(\alpha)$. The 25 percent trimmed mean $m(0.25)$ is called the *midmean*, and $m^c(0.25)$ is the *outmean*. A *Winsorized mean* is the mean of a Winsorized sample. A sum of squares of all the observations around the Winsorized mean is a *Winsorized sum of squares*. A t-statistic using a Winsorized sum of squares in the denominator, and a trimmed mean in the numerator is a *trimmed-t*, while using a Winsorized mean in the numerator results in a *Winsorized-t*. In each case there is a necessary adjustment in degrees of freedom.

It sometimes happens that the exact magnitude of some sample values are not known: The observer knows only that those larger than (or smaller than) some point are unavailable. A sample trimmed in this way is said to be *censored*. The sample is singly censored if the unavailable observations are at one end and *doubly censored* if they occur on both ends of the sample. In *Type I censoring* the *point of censoring* is fixed in advance and known to the observer. Consequently, the number of points to be censored is a random variable not known in advance. In *Type II censoring* a fixed percentage of the sample is censored, and the *number* of censored observations are known in advance. As a consequence, the point of censoring is a random variable. *Truncation* differs from censoring in that the population rather than the sample is restricted in one or both tails.

The study of extreme phenomena (floods, droughts, earthquakes, breaking strengths, etc.) has some terminology of its own. The ,extremal quotient is $|X_{(n)}/X_{(1)}|$. The *m-th distance from the top* is $i_m = X_{(n-m+1)} - X_{(n-m)}$. The *return period* of a value equal to or larger than x is the function $T(x) = 1/(1 - F(x))$, and the *intensity function* is $\mu(x) = f(x)/(1 - F(x))$. The *characteristic largest and smallest values*, u_n and u_1, are the quantiles defined by $F(u_n) = 1 - 1/n$ and $F(u_1) = 1/n$, respectively. Evaluating the intensity function at the 2 characteristic extreme values, u_n and u_1, yields the *external intensities* α_n and α_1. The quantities $y_n = \alpha_n (X_{(n)} - u_n)$ and $y_1 = \alpha_1(X_{(1)} - u_1)$ are called *reduced extremes*. For symmetric distributions $\alpha_1 = \alpha_n$, and the *reduced range* is then the difference between the reduced extremes. The *m-th extreme* is the *m*-th value from the top or bottom of the sample, and the quantiles u_m and $_mu$ defined by $F(u_m) = 1-m/n$ and $F(_mu) = m/n$ are called the *characteristic m-th largest and the m-th smallest* values or *characteristic m-th extreme* values, while $y_m = \alpha_m(x - u_m)$ and $y_m = {_m}\alpha(X - {_m}u)$ are the *reduced m-th extremes*. Extreme value theory deals primarily with the asymptotic distribution of extremes.

Order statistics are also used in slippage tests. Given a sample of size *m* from each of *n* continuous populations, a *slippage test* is used to decide whether one of the populations (we do not know which) has slipped to the right (or left) of the others in either the mean or the variance.

Outliers (mavericks, sports, flyers, or *wild, anomalous,* or *discordant* observations) are observations so far from the bulk of the data in a sample that they *appear* surprising or discrepant to the observer. In making that subjective judgment, the observer has a *model* in mind and believes the outliers depart from the model. To know what to do with them, he must first decide whether the objective is *estimation* (of the mean and variance) or whether the outliers are interesting in themselves. In prospecting for uranium ore, for example, he is not at all interested in estimating the background; instead he is interested in large outliers that indicate a deposit. If estimation is the goal, the usual model is normality of the bulk of the data. If he can *detect* which observations are not part of the underlying normal population, he can delete those observations and proceed to do his estimation on the remainder of the sample. Another solution is to *accommodate* the outliers by means of a robust estimator. That usually means that the outliers are given low weights in the estimation. Even with detection, the aberrant observations should be recorded and the reason for deleting them explained. They may provide a vital clue to something going amiss in the data-gathering procedure. In other cases the outliers contain useful information or serendipitous discoveries (as was the case with penicillin). For detecting one outlier from a normal population, the extreme studentized deviate is used as a statistic. If multiple outliers are suspected, single outlier tests are likely to fail. If 2 large outliers are close together, there is a *masking* effect, which "hides" one of them. In that case a 1-outlier test cannot get started. One the other hand, if there is only 1 outlier and it is quite large, a 2-outlier test is likely to declare 2 outliers, a phenomenon called *swamping*. The best procedures for multiple outliers require that the user specify an upper limit *k* on the number of possible outliers and work backward (check the largest *k* observations one at a time, beginning with the 1 closest to the mean of the bulk of the data) until an observation is decided to be contaminated. (All those further out are then judged to be contaminated.) Multiple outliers from a linear model or designed experiment are harder to detect, and research is still going on in that area.

References

The best in this field is David, H. A. 1981. *Order Statistics*. New York: Wiley and Sons. There is a large literature on outliers, one good reference being Barnett, V., and Lewis, T. 1978. *Outliers in Statistical Data*. New York: Wiley and Sons.

9

Stochastic Processes

We read that in a certain country the birth rate per thousand population is 34. What does that mean? It does not mean that for every randomly chosen group of 1,000 persons there will be exactly 34 births per year; instead, there is an *average* of 34 births per thousand persons per year. For any particular group the actual number of births fluctuates about 34 (the birth rate is a conditional probability). It is implied that the birth "process" remains rather steady over time, a property that we call *stationarity*. Birth and death processes, nuclear reactions, wind velocities, Brownian motion, random walks, and shot noise in tubes are just a few of the processes that vary with time and that we describe in this chapter.

A *stochastic process* $X(t)$ is a family of random variables indexed by (depending on) a parameter t, which runs over an index set T. The parameter t usually denotes "time," and at any specific time t, $X(t)$ is a random variable. $X(t)$ is also called a *random process* or simply a *process*. The index set T is called the *parameter set*. If T is countable, $X(t)$ is a *discrete* (or *discrete time* or *discrete parameter*) process or a *stochastic sequence*. If T is an interval, finite or infinite, $X(t)$ is said to be a *continuous time* or *continuous parameter* process. The index set is not necessarily one dimensional. For example, $X(t)$

may represent the height of a wave at a point t, which has latitudinal and longitudinal coordinates.

At a fixed time t, $X(t)$ is a random variable with a set of possible outcomes, each of which is called a *state* s. The set of all possible outcomes is the *state space*. $X(t)$ can thus be thought of as a function of t and s. If $X(t)$ is assigned a particular value or state for each t, the resulting function of t is a *sample function* or a *realization* of the process. The collection of all possible realizations is an *ensemble*. There is a *mean value function* defined for each t by $m(t) = E[X(t)]$. The *autocovariance* or *covariance kernel* of $X(t)$ is $K(t_1, t_2) = \text{Cov}[X(t_1), X(t_2)] = E[(X(t_1) - m(t_1)) (X(t_2)) - m(t_2))]$ for each fixed value of t_1 and t_2. The *variance* of $X(t)$ is $K(t,t) = \sigma^2(t)$, and the *autocorrelation* of $X(t)$ is $R(t_1,t_2) = K(t_1,t_2)/\sigma(t_1)\sigma(t_2)$. Two processes $X(t)$ and $Y(t)$ are *uncorrelated* if $\text{Cov}[X(t_1),Y(t_2)] = 0$ for all t_1 and t_2. They are *orthogonal* if the group $X(t_1)$, $X(t_2)$, . . . , $X(t_n)$ is independent of $Y(t'_1)$, . . . , $Y(t'_m)$ for all t_1, . . . , t_n, t'_1, . . . , t'_m.

The functions $m(t)$ and $K(t_1 t_2)$ are examples of *ensemble averages* or expectations. There are corresponding *sample averages*, such as the *sample mean* and *sample covariance*. For any function $g[X(t)]$ of a discrete parameter process, the ensemble average is $E(g(X(t))]$, and the sample averages are $1/T \sum_{t=1}^{T} g(X(t))$. For a continuous parameter process, there is a *time average*

$1/T \int_0^T X(t)dt$ or $1/2T \int_{-T}^{T} X(t dt$, where the integral is defined in the Riemann sense and the limiting sum converges in the mean square sense. A stochastic process is *ergodic* if the sample (or time) averages converge to the ensemble or population averages. An *ergodic theorem* gives conditions under which the sample averages converge to the corresponding ensemble averages.

If the state space S is countable, S is a *discrete state space*; otherwise S is a *continuous state space*. $X(t)$ is a *real-valued process* if it assumes only real values. If S consists only of integers, $X(t)$ is an *integer-valued process*. If $X(t)$ assumes values in Euclidean k-space, $X(t)$ is a *k-vector process*.

Two important properties used in classifying real-valued stochastic processes are the following: (1) the process has *independent increments* if $X(0) = 0$ and if the *increments* $Z_1 = X(t_2) - X(t_1)$, $Z_2 = X(t_3) - X(t_2)$, . . . , $Z_{n-1} = X(t_n) - X(t_{n-1})$ are mutually independent for all $t_1 < t_2 < . . . < t_n$ and (2) the process is *strictly stationary, strongly stationary*, or simply *stationary* if the joint distribution of $[X(t_1), . . . , X(t_n)]$ is the same as the joint distribution of $[X(t_1 + h), . . . , X(t_n + h)]$ for all $h > 0$ and aribtrary $t_1, t_2, . . . , t_n$. In other words, the distribution of the process is independent of time (there are no trends in the means of the variances). The process is *weakly stationary* or *wide-sense stationary* or *covariance stationary* if the second moments are finite, the mean is a constant, and $\text{Cov}[X(t), X(s)]$ depends only on $|t - s|$. Strictly stationary processes are so important because they have constant mean and variance (if the second moments are finite), and because the processes are ergodic. Nonstationary processes are said to be *evolutionary*.

We now consider the rich field of Markov processes. A process has the *Markov property* or is a *Markov process* if future probabilities of the process are uniquely determined by the *present* stage of the process. In other words, the future of the process does not depend on past values. We write that condition as $P[X(t_{n+1}) \leq x_{n+1}|X(t_1) = x_1, \ldots, X(t_n) = x_n] = P[X(t_{n+1}) \leq x_{n+1}|X(t_n) = x_n]$. A 1-dimensional random walk is an example. Related to the idea of a Markov process is a *martingale*, whose conditional expectation of the "next" observation is equal to the present value of the process, i.e., for every n and $t_1 < \ldots < t_{n+1}$, $E[X(t_{n+1})|X(t_1) = x_1, \ldots, X(t_n) = x_n] = x_n$ for all $x_1, \ldots, x_n$. As an example, let X_n represent a gambler's fortune on the n-th throw of the dice. If his fortune on the next throw is equal (on the average) to his present fortune, the process is a martingale, and in that sense it is a "fair" game.

We now consider 2 continuous time processes that play a key role in Stochastic Processes: the Weiner and Poisson processes. As usual, normal random variables play an important part. A process $X(t)$ is *normal* if the random variables $X(t_1), \ldots, X(t_n)$ are jointly normal for any n and $t_1, \ldots, t_n$. A Markov process, $X(t)$, $t \geq 0$, is a *Weiner* or *Weiner-Levy* or *Brownian Motion process* if $X(t)$ has stationary, independent increments and if, for every $t > 0$, $X(t)$ is normally distributed with $E(X(t)) = 0$ and $X(0) = 0$. That implies that the increments $X(t) - X(s)$ are normal with mean zero and variance $\sigma^2|t - s|$. The process was so named because a small particle immersed in a liquid or gas shows a ceaseless irregular motion resulting from continual bombardment of molecules in the medium. That intensively studied motion became known as *Brownian Motion*. When $X(t)$ is a Weiner process, $Z(t) = \int_0^t X(s)ds$ is an *integrated Weiner process*. A normal process with zero mean and $\text{Cov}[X(u), X(t)] = \alpha \exp(-\beta|u - t|)$ for $\alpha, \beta > 0$ is an *Ornstein-Uhlenbeck* process.

Processes whose realizations are made up of points in time are called *point processes*. A very important process associated with a point process is the integer-valued *counting process*, $N(t)$, $t \geq 0$, which counts the number of points occurring in the time interval $(0,t)$. Well-known examples of counting processes are those that count the number of disintegrations emitted during radioactive decay, the number of phone calls arriving at a switchboard, and the number of breakdowns of a machine. One particular counting process that occupies an important place is the (*homogenous*) *Poisson process*, which satisfies 5 axioms: (1) $N(0) = 0$, (2) the increments are independent, (3) they are stationary, (4) in any interval, however small, there is a positive probability that an event will occur, but an event is not certain to occur, and (5) it is not possible for events to occur simultaneously. Karlin has given an entertaining and enlightening example: Let $N(t)$ denote the number of fish caught in the interval $(0, t]$. If the number of fish in the stream is very large and if everyone stands an equal chance of catching fish and if they are as likely to nibble at one instant as another, $N(t)$ is a Poisson process. It has the Markov property because the chance of catching a fish does not depend on the number already caught. The most characteristic property is that there is no premium for waiting: The fisherman who has just arrived has the same chance of catching a fish as the fisherman who has been waiting for hours. For any fixed t, $N(t)$ has a Poisson distribution with parameter vt, i.e., $m(t)$ is directly proportional to t, the constant of proportionality being equal to v, the mean rate, or intensity of the counts. If $m(t)$ is a function of t, rather than a constant times t, then $N(t)$ is a *nonhomogeneous Poisson process* with intensity function $v(t)$, and $m(t) = \int_0^t v(s)ds$. Also, the increments are nonstationary. In case $N(t)$ satisfies all the axioms except that events *can* happen simultaneously, it is a *generalized Poisson process*. A process that takes on only 2 values is called a *dot-dash process*, a *1-minus-1 process* (if $-1,1$ are the values), or a *zero-1 process* (if $0,1$ are the 2 values). If $N(t)$ counts the number of times a 1-minus 1 process $X(t)$ changes value in $(0,t]$ and if $N(t)$ is a Poisson process, then $X(t)$ is a *random telegraph signal*.

Let $\{Y_n, n = 1, 2, \ldots\}$ be a family of r.v.'s independent and identically distributed as the r.v. Y. Let $N(t)$ be a Poisson process independent of the sequence $\{Y_n\}$. Then the random sum of random variables $X(t) = \sum_{n=1}^{N(t)} Y_n$ is a *compound Poisson process*, and it has stationary independent increments. If the r.v. Y is integer-valued, then $X(t)$ is also a generalized Poisson process. Any generalized Poisson process can be represented as a compound Poisson process.

A filtered Poisson process is one that arises from linear operations on a Poisson process, and those processes first served as models of shot noise. Specifically, a stochastic process $\{X(t), t > 0\}$ is a *filtered Poisson process* if it can be represented as $X(t) = \sum_{m=1}^{N(t)} w(t, \tau_m, Y_m)$, where $N(t)$ is a Poisson process, Y_n is a sequence of i.i.d. random variables independent of $N(t)$, and $w(t, \tau, y)$, a function of 3 variables, called the *response function*. $X(t)$ is thus the value at time t of the sum of $N(t)$ signals arising in the interval $(0,t)$, where $w(t, \tau_m, y)$ is the value at time t of a signal of magnitude y originating at time τ_m.

A stochastic process $\{X(t), -\infty < t < \infty\}$ that can be represented as the superposition of impulses occurring at random times $\ldots \tau_{-1}, \tau_0, \tau_1, \ldots$ is called a *shot noise process*. The impulses are assumed to have the same shape

$w(s)$ so that the process can be represented as $X(t) = \sum_{-\infty}^{\infty} w(t - \tau_m)$. *Campbell's Theorem* states that $E[X(t)] = \nu \int_{-\infty}^{\infty} w(s)ds$ and $\mathrm{Cov}[X(t),X(t+v)] = \nu\int_{-\infty}^{\infty} w(s)w(s+v)ds$. A shot noise process is a filtered Poisson process.

Given events occurring in the interval $(0,\infty)$ we define T_1 as the time from 0 to the first event, T_2 as the time from the first event to the second, etc. The T_i are the *inter-arrival times*. The variable $W_n = T_1 + T_2 +, \ldots, + T_n$ is the *waiting time* until the n-th event occurs. For a Poisson process with intensity v, W_n has a gamma distribution with parameters n and v, and the inter-arrival times are independent and have an exponential distribution with parameter $1/v$. Given that exactly K events occur in $(0,T)$, the times at which the events occur are independent and uniformly distributed over $(0,T)$.

As example of a continuous time process that is, in general, non-Markovian, is the renewal process. Consider an object such as a light bulb. When the life of the bulb is ended, it is immediately replaced by a new bulb. We then wish to know the distribution of the total number of replacements in a given time t. A *renewal process* or *renewal counting process* is a continuous parameter, nonnegative, integer-valued counting process $N(t)$, which registers the number of events (renewals) in the interval $(0,t]$ when the inter-arrival times are independent and identically distributed positive random variables. In case the inter-arrival times are exponentially distributed, $N(t)$ is a Poisson process. The expected value of $N(t)$ is the *renewal function*. The distribution function of a renewal process, $F(x)$, is zero for $x < 0$ and continuous at $x = 0$. If $F(x)$ has a jump q at $x = 0$, $N(t)$ is a *modified renewal process*. If the inter-arrival times are independent and all except the first are identically distributed, $N(t)$ is a *delayed renewal process*.

If $f(t)$, $g(t)$, and $h(t)$ are functions defined on $t \geq 0$ that satisfy an integral equation of the type $g(t) = h(t) + \int_0^t g(t-s)f(s)ds$, the equation is called a *renewal equation*. The mean value function $m(t) = E(N(t))$ of a renewal process satisfies the renewal equation $m(t) = F(t) + \int_0^t m(t-s)f(s)ds$, where $F(x)$ and $f(x)$ are the cdf and pdf of the inter-arrival times.

The time from the last renewal up to time t is the *current life* or *age* of the unit. The *excess life* $\gamma(t)$ is the amount of life remaining in the unit. The sum of the current life and excess life is the *total life*. If $g(t,x) = P[\gamma|t] > x$, then $g(t,x)$ satisfies the renewal equation $g(t,x) = 1 - F(t+x) + \int_0^t g(t-s)f(s)ds$. For exponential inter-arrival times, excess life is exponentially distributed with the same mean as the inter-arrival times. If the inter-arrival times are distributed as the r.v. T and if $\mu = E(T)$ and $\sigma^2 = \mathrm{Var}(T)$ are finite, $N(t)$ is asymptotically normal with mean equal to μ/t and variance equal to $t\sigma^2/\mu^3$.

A general Markov process is described by its *transition probability function*, $P(E,t|x,t_0)$, which is the conditional probability that at time t, $X(t)$ belongs to the set E given that at time t_0 the system was in state x. The process is *homogeneous in time* or has *stationary transition probabilities* if $P(E,t|x,t_0)$ depends only on the difference $(t - t_0)$. A Markov process with a discrete state space is called a *Markov chain*, and a great many physical, biological, and economic processes can be modeled as Markov chains.

For Markov chains, the states are customarily labeled by the nonnegative integers $(0, 1, 2, \ldots)$. For a discrete parameter Markov chain we write X_n instead of $X(n)$, and X_n may be thought of as the outcome of the n-th "trial." If the state is finite, the Markov chain is finite. The probability that the outcome of the $(n + 1)$st trial is equal to j given that the outcome of the n-th trial was equal to i is denoted by $p_{ij}\,(n,n + 1)$ and called a *one-step transitional probability*. The probability that the outcome of the $(n + m)$th trial is j given that the outcome of the m-th trial was i is the *n-step transitional probability* $p_{ij}(m,n + m)$. If $p_{jk}(m,n)$ depends only on the difference $(n - m)$, the Markov chain is *homogeneous* or has *stationary transition probabilities*.

For a homogeneous Markov chain, we denote $p_{ij}(m,n + m)$ by $p_{ij}(n)$ and the one-step probability $p_{ij}(1)$ by p_{ij}. For continuous parameter Markov processes $X(t)$, $t \geqslant 0$, we can also write $p_{jk}(s,t) = P[X(t) = k|X(s) = j]$ and call $p_{jk}(s,t)$ the *transition probability function*. In either case the probabilities satisfy the *Chapman-Kolmogorov Equation* $p_{jk}(m,n) = \Sigma\ p_{ji}(m,u)\ p_{ik}(u,n)$, where s and t can be substituted for m and n. The equation states that the move from state j to state k in time $(n - m)$, or in $(n - m)$ trials, is accomplished by moving from j to j' in $(u - m)$ trials and then from j' to k in $(n - u)$ trials.

A *transition probability matrix* $P(m,n) = (p_{ij}(m,n))$ of transition probabilities can be written down, and the Chapman Kolmogorov Equations then state that $P(m,n) = P(m,u)P(u,n)$. The probability law of a Markov chain is completely determined by knowledge of the transition probability matrix and the initial conditions.

The rows of a transition probability matrix sum to 1, and any matrix with that property is called a *Markov* or *stochastic matrix*. If the column sums are also equal to 1, the matrix is *doubly stochastic*.

A useful type of Markov Chain, encountered in other areas as well, is the random walk, so called because it aptly describes the path of a drunk moving randomly 1 step forward or backward. The position of a moving "particle" is frequently used to describe the state of the system. A (one-dimensional) *random walk* is a discrete-time Markov chain in which the particle in state i can only stay in state i or move 1 unit to the right or left. If the probability of a move in either direction is equal to one-half, the random walk is *symmetric*. If the particle cannot move at all, it has reached an *absorbing barrier*, and if it reaches a state that automatically returns it to the previous state, it has reached a *reflecting barrier*. Both barriers are special cases of the *elastic barrier*, which has probability p of moving from state 1 to 2, probability $\delta(1-p)$ of staying in state 1, and probability $(1-\delta)(1-p)$ of moving from state 1 to state 0 and being absorbed there.

One classic random walk is referred to as the *gambler's ruin*. Suppose gamblers A and B start off with a and b dollars respectively, and at each turn A wins 1 dollar with probability p and loses 1 with probability $1-p$. Let $X(n)$ be B's gain after n turns. If $X(n)=a$, A is ruined. A classic problem is to find the probability of ruin for each contestant, given a, b, and p.

For a Markov Chain, the number of times the state k is visited in the first n transitions is the *occupation time* (or *sojourn time*) for state k; the limit (as n gets large) of the occupation time is the *total occupation time* of state k. If the system is in state i, and T_i is the first time the system leaves state i, T_i is called the *waiting time* in state i. Given that the system is in state i, the time to first entrance into state j is called the *first passage* (or *first entrance*) *time* from i to j. The waiting times, occupation times, first passage times, and absorption times are all random variables with associated probability distributions.

However, the transition probability matrix can be decomposed in several ways that require a classification of the states of a Markov chain. State j is *accessible* from state i if there is a positive probability that j can be reached from i in a finite number of steps. If states i and j are accessible to each other, they are said to *communicate*. A Markov Chain is *irreducible* or *indecomposable* if every state can be reached from every other state, i.e., if all states communicate with each other. A state that communicates with itself is a *return state*; one that communicates with no other states is called a *nonreturn state*. For any state j, its *communicating class* $C(j)$ is the set of states that communicate with j. A nonempty set of states C is *closed* if no state outside C is accessible from a state in C, i.e., starting in C, we stay in C. A set that is not closed is *nonclosed*. Once a Markov chain enters a closed class, it remains there.

Communication of states is an equivalence relation, so that the totality of states can be partitioned into equivalence classes within which the states communicate; it is possible to get from 1 class into another, but not back. A state j is *periodic* (with period t) if return to j is impossible except at times t, $2t$, $3t$, . . . , and t is the greatest integer larger than 1 with that property. A state is *aperiodic* if it has period 1. The chain is *aperiodic* if each state is aperiodic. Two states that communicate with each other have the same period; hence the *period of a communicating class* is the period common to the states in the class. A state is *essential* if it communicates with every state accessible from it; a state not having that property is *unessential*. A state k is *recurrent* (or *persistent*) if a process that starts from k is certain to return to k; it is *nonrecurrent* (or *transient*) if the return to k is not certain. A class of states of a Markov chain is *recurrent* (nonrecurrent) if every state in the class or chain is recurrent (nonrecurrent). A communicating class is either recurrent or nonrecurrent, and a recurrent communicating class is closed.

The time required for a first return from state j to state j is the *recurrence time* for state j, and its expectation is the *mean recurrence time*. A recurrent state is *positive* if the mean recurrence time is finite and *null* if it is infinite. A recurrent state that is neither null nor periodic is by some authors called *persistent* or *ergodic*. A process is called current, ergodic, positive, null, or transient if every one of its states has the corresponding property.

We speak of 2 *decompositions of a Markov Chain*: (1) into a finite or countable disjoint collection of states in which each set of states is either a communicating class or is a single nonreturn state, and (2) into a union of disjoint communicating classes in which each class is either closed and recurrent, closed and nonrecurrent, or nonclosed and nonrecurrent. The second decomposition is a strengthening of the first. The decomposition is a result of the fact that recurrent communicating classes are closed, and closed nonrecurrent communicating classes have an infinite number of states.

For continuous parameter Markov chains the part of the one-step transition probabilities is taken by *transition intensities*, which are the derivatives of transition probability functions evaluated at zero. The limit (as $h \rightarrow 0$) of $p_{jk}(t, t+h)/h$ is called the *intensity of transition* to k, given that the chain was in state j at time t. The limit of $(1 - p_{jj}(t, t+h))/h$ is the *intensity of passage* given that the chain was in state j at time t. If $\Sigma p_{jk}(s,t) < 1$, the Markov process is said to be *dishonest* or *pathological*. If $p_{ij}(n)$ are the n-step transitional probabilities for a homogeneous Markov Chain, let π_{ij} be the limit, as n gets large, of $\sum_{k=1}^{n} p_{ij}(k)/n$. The limit exists, and the chain is *dissipative* if $\pi_{ij} = 0$ for all i and j, *semidissipative* if $\pi_{ij} = 0$ for some i and j but $10\sum_{j=i}^{\infty} \pi_{ij} < 1$, *nondissipative* if $\sum_{j=1}^{\nu} \pi_{ij} = 1$ for all i. A state j is *positive* when $\pi_{jj} > 0$ and *dissipative* when $\pi_{jj} = 0$.

A *birth and death process* is a continuous parameter-integer-valued counting process $N(t), t \geq 0$, with transitions taking place only from one state to an immediate neighbor (making it the continuous-time analog of a random walk). $N(t)$ represents the size of a population at time t, and in a small interval of time the population either stays the same, increases by 1 unit (a birth), or decreaseds by 1 unit (a death). The probability of a given individual increasing by 1 unit is essentially proportional to the length of the interval, i.e., $\lambda \Delta t$. The conditional probability of an increase by 1 unit is denoted by $\lambda_n(t)$ called the *birth rate* (*parameter*). The probability of a decrease by 1 unit is denoted by $\mu_n(t)$, called the *death rate*. Those parameters may depend upon the time

t at which the process is observed and the population size n at the time. If the parameters do not depend upon t, the process is said to be *homogeneous*. If $\mu_n(t)$ is zero, it is called a *pure birth process*, and if $\lambda_n(t)$ is zero, it is labeled a *pure death process*. If the birth and death rate parameters are directly proportional to the population size n, we have *linear* (or *simple*) birth and death rates. The homogeneous birth process with linear birth rate is referred to as the *Yule* (or *Furry*) *process*. If $\lambda_n = \lambda n$, $\mu_n = \mu n$, the process is a *linear growth process*. If $\lambda_n = \lambda n + a$, $\mu_n = \mu n$, the process is one with *linear growth with immigration*. If $\lambda_n = v$ is constant, the process is a Poisson process with intensity v. The *Polya process* is a nonstationary or nonhomogeneous pure birth process in which λ_n depends on time: $\lambda_n(t) = \dfrac{1 + an}{1 + at}$. The *Galton-Watson* process is a binary birth process with birth rate $\lambda > 0$ such that the population size doubles at intervals of time equal to $1/\lambda$. The *Ehrenfest process* is a finite birth and death process with $\lambda_n = (N - n)p$, $\mu_n = n(1 - p)$, $0 \leq n \leq N$, and p is the probability of a birth. If the size $X(t)$ of a population varies between 2 fixed integers N_1 and N_2, and if the birth and death rate per individual at time t are linear functions of the instantaneous population size, the process is a *logistic process*.

Epidemics are represented as nonlinear death processes. If infected individuals are removed from the population by death or isolation, the process is known as a *Bartlett-McKendrick process*.

Consider a population of size $X(0)$. Let each individual in the population give birth at his death (or split into) a random number k of new individuals of the same type with probability p_k, where $p_k \geq 0$ and $\Sigma p_k = 1$. Direct descendants of the $(n-1)$st generation from the n-th generation, and $X(n)$ denotes the size of the population in the n-th generation. The individuals in each generation act independently of each other. The sequence $\{X(n)\}$ is an integer-valued Markov chain called a (discrete) *branching* (or *cascade* or *multiplicative*) *process*. If an individual can given birth to N different types of progeny, the branching process is *N-dimensional*.

A branching process that takes account of age structure is said to be *age-dependent*. An example is the *Bellman-Harris Model*. That type of branching process is non-Markovian and belongs to the class of *regenerative processes* in which, for some particular time T and thereafter, the conditional distribution of $X(t)$ given $X(T)$ equals the conditional distribution of $X(t)$ given $X(\tau)$ for all $\tau \leq T$. T is called the *regeneration point*.

An important problem in branching processes is to find the limiting *probability of extinction* of the population. Given that $X(0) = 1$, the *fundamental theorem of branching processes* gives that probability (p) as the smallest root of the equation $P(p) = p$, where $P(z)$ is the generating function of the number of offspring per individual. A *generalized discrete branching process* is described by a Markov matrix in which the initial state is absorbing and there is a positive probability that a population of i individuals does not reproduce.

Another type of Markov chain is typified by a box office with a single cashier. The arrival times of the customers have a Poisson distribution, while the service times are independently and identically distributed. If $X(n)$ represents the number of persons waiting to be served at the moment the n-th customer departs, $X(n)$ is said to be an *imbedded* Markov chain, meaning that it arises from a more basic Markov chain.

A general *queueing process* describes a "waiting-line" situation. The "customers" (which might be phone calls, broken machines, ships, etc.) arrive at random times at the facility and request service of some kind. The service is restricted in some manner, and the customers are required to form a "queue" to wait for service. The purpose of analyzing the queue is to improve the service (cut back on waiting time). The 3 main aspects of the process are (1) the *input process*, which is described by the distribution of arrival times of the customers, (2) the *service mechanism*, which is specified by the number of servers and the distribution of service times, and (3) the *queue discipline*, which describes the way in which the waiting line is formed, maintained, and dealt with. Queue length, waiting times, and first passage times are the usual problems. *Arrival times* are the times when the customers arrive. *Inter-arrival times* are the elapsed times between arrivals. The *traffic intensity* is the ratio of expected length of service time per customer to the expected length of inter-arrival time. The *virtual waiting time* is the time a customer would have waited for service had he arrived at time t. The *busy period* is the time interval during which the server is continuously busy. For telephone queues the service time (duration of conversation) of the n-th call is the *holding time*.

References

Stochastic processes is not an elementary subject, and I do not know of any easy book. For a first reading I would use Bailey, N. T. 1964. *The Elements of Stochastic Processes*. New York: Wiley and Sons. or Parzen, E. 1962. *Stochastic Processes*. San Francisco: Holden-Day. The classic in this field is Karlin, S., and Taylor, H. M. 1968. *A First Course in Stochastic Processes*. New York: Academic Press. For an easier book on Markov chains, I recommend Kemeny, J. G., and Snell, J. L. 1960. *Finite Markov Chains*. Princeton, N.J.: Van Nostrand.

10

Time Series

A *time series* is a realization of a stochastic process; it is a set of observations generated sequentially in time. If the set is continuous (e.g., a continuous record of wind velocity), the time series is *continuous* and will be denoted by $x(t)$. If the set is finite (e.g., daily stock market averages), the time series is *discrete* and will be denoted by x_t. Discrete time series frequently arise from sampling a continuous time series.

If we record wind velocity and wind direction at the same instant of time, we have a *multivariate time series*, specifically a bivariate one. If we record temperature of a heat source at time t_1 and simultaneously note the (increasing) distance t_2 from the heat source, we have a *multidimensional time series*, specifically a 2-dimensional one. In the first case there are 2 series measured on a common basis; in the second case, 1 series with 2 components.

The object of a time series analysis is (1) to describe the series, (2) to explain or model its behavior, and (3) to use the model to forecast or control the future behavior of the series. The analysis must take into account the fact that neighboring values of the series are correlated. Methods that require independence are not applicable.

Stationary stochastic processes have a constant mean and constant variance, and the probability distribution is the same for all times t. The shape of the distribution can thus be inferred by forming a histogram of the time series $x_1, x_2, \ldots x_N$, and the mean μ and variance σ^2 of the process can be estimated from $x = \Sigma x_i/N$ and $\hat{\sigma}^2 = \Sigma(x_i - \bar{x})^2/N$. Stationarity (being stationary) also implies that the joint distribution is the same for all t_1 and t_2 that are the same distance apart in time. The process covariance between X_t and X_{t+k} is called the *autocovariance at lag k*, denoted by

$$\gamma_k = E[(X_t - \mu)(X_{t+k} - \mu)].$$

A plot of γ_k versus k is the *autocovariance function*. The *autocorrelation at lag k is* $\rho_k = \gamma_k/\gamma_o$, (also called the *serial correlation of order k*), and a plot of ρ_k versus k is called the *autocorrelation function*, a term that is replacing *correlogram*. The best estimate of γ_k is made from the time series: $c_k =$
$$\sum_{i=1}^{N-k} (x_t - \bar{x})(x_{t+k} - \bar{x})/N.$$

Given a function $f(t)$, where t is a real variable, the function $F(\omega) = \int_{-\infty}^{\infty} f(t)e^{-i\omega t}dt$ is the *Fourier integral* or *Fourier transform* of $f(t)$. The *inversion formula* allows us to represent $f(t)$ in terms of its Fourier transform: $f(t) = \frac{1}{2\pi} \int_{-\infty}^{\infty} F(\omega)e^{i\omega t}d\omega$. Given 2 functions, $f(t)$ and $g(t)$, the integral $h(x) = \int_{-\infty}^{\infty} f(y)g(x - y)dy$ is called the *convolution* of $f(t)$ and $g(t)$.

Time-invariant linear differential equations with constant coefficients always have solutions $h(x)$ that are convolution integrals with $g(x)$ being called the *forcing function*. The *output* $h(x)$ can also be thought of as a weighted sum of past values of the *input* $f(y)$ and the *weight* or *weighting function* $g(y)$. The operation of convolution is frequently written as $h(x) = f(x)*g(x)$. The *time convolution theorem* states that the Fourier transform of the convolution of 2 functions is equal to the product of the Fourier transforms of the 2 functions. The *frequency convolution theorem* states that the Fourier transform of the product of 2 functions equals the convolution of their Fourier transforms. Given that $F(\omega) = A(\omega)e^{i\phi(\omega)}$ is the Fourier transform of $f(t)$, *Parseval's Theorem* (or *formula*) states that $\int_{-\infty}^{\infty} |f(t)|^2 dt = \frac{1}{2\pi} \int_{-\infty}^{\infty} A^2(\omega)d\omega$.

Linear filters are useful tools in smoothing and modeling time series. A moving average is a linear filter used for smoothing. The earth acts as a linear filter for seismic disturbances. A *linear filter L* transforms a time series input $x(t)$ into a time series output $y(t)$. L has the *time invariance property* that states that $L(x(t + h)) = y(t + h)$ whenever $L(x(t)) = y(t)$ and the *linearity property*, which states that $L(\alpha x(t) + \beta z(t)) = \alpha L(x(t)) + \beta L(z(t))$. For a continuous input $x(t)$, the linear filter can be represented as a convolution integral $L(x(t)) = \int_{-\infty}^{\infty} h(u)x(t - u)du$. If the Dirac delta function $\delta(t)$, which is a spike or impulse, is used as an input, the result is $L(\delta(t)) = h(t)$, so that $h(t)$ is called the *impulse response function* of the filter. Since an arbitrary input is representable as a sum of impulse functions, the output will be a sum of impulse response functions; hence the impulse response function completely determines the properties of the filter. Arbitrary inputs can also be represented as a sum of step functions or as a sum of Fourier components (periodic functions), hence the *step response function* and the response to $e^{i\lambda t}$ completely determine the properties of the filter. In fact, a linear filter transforms $e^{i\lambda t}$ into $B(\lambda)e^{i\lambda t}$, where $B(\lambda)$ is called the *transfer function* of the filter. The *transfer function* or *system function* or *frequency response function* is the Fourier transform of the impulse response function. The transfer function thus determines the properties of the filter.

A linear filter is *stable* if its response to a bounded output is bounded; it is *realizable* when the output at time t depends only on the input at times $s \leq t$. Filters that attenuate high frequencies and pass low frequencies with little change are called *low-pass filters*. *High-pass filters* attenuate low frequencies and pass high ones. A low-pass and a high-pass filter can be combined to remove all but a band of frequencies; such a filter is called a *band-pass* filter. *Ideal low-pass* (high-pass) *filters* completely eliminate the power above (below) a certain *cut-off frequency* and leave untouched the frequencies below (above) it. Linear combinations of linear filters are linear filters. Those properties allow us to design filters that will produce a given output.

A *digital filter* is a linear filter in discrete time. By analogy with the convolution integral, it can be represented as a sum $L(x_t) = \sum_{-\infty}^{\infty} c_j x_{t-j}$, where the c_j are real numbers called the *filter weights* and are such that $\Sigma\, c_j^2 < \infty$. For a time series $x(t)$ representable by periodic functions, $x(t) = \Sigma_\lambda \alpha_\lambda e^{i\lambda t} = \Sigma |\alpha_\lambda| e^{i(\lambda t + \psi(\lambda))}$, where $|\alpha_\lambda|$ is the *amplitude* and $\psi(\lambda)$ the *phase angle* of α_λ. If we write $B(\lambda) = |B(\lambda)| e^{i\theta(\lambda)}$, $|B(\lambda)|$ and $\theta(\lambda)$ are called the *gain* and *phase* (shift) functions of the filter respectively. A plot of log gain versus log frequency and phase versus log frequency is called a *Bode-plot* of the filter.

The most popular approach to *modeling* a time series is to use the autoregressive and moving average models, sometimes referred to as *Box-Jenkins models*. Consider a sequence of independent random variables $a_t, a_{t-1}, \ldots$ that are normally distributed with zero mean and variance σ_a^2. The sequence $\{a_i\}$ is called a *white noise process*, and the a_i are referred to as *shocks*. A linear filter transforms a white noise process into a process whose current value, z_t, is a weighted sum of the current and previous shocks: $z_t = \mu + a_t + \psi_1 a_{t-1} + \psi_2 a_{t-2} + \ldots$. If the sequence $\{\psi_i\}$ is finite or convergent, the process z_t is *stationary* and the linear filter is *stable*. Let $\tilde{z}_t$ denote the deviation $z_t - \mu$. An *autoregressive model* is a linear combination of previous values of the $\tilde{z}_t$ plus a current shock. In particular, $\tilde{z}_t = \phi_1 \tilde{z}_{t-1} + \phi_{t-2} + \ldots + \phi_p \tilde{z}_{t-p} + a_t$ is an *autoregressive process of order p*, $(AR(p))$. The name is applied because it resembles a regression of $\tilde{z}_t$ on previous values of the same process. The process has a "memory," since it systematically depends upon past history but includes a disturbance happening at time t. If $\tilde{z}_t$ is a finite linear combination of the previous and present shocks, then z_t is a finite *moving average process*. In particular, $\tilde{z}_t = a_t - \theta_1 a_{t-1} - \theta_2 a_{t-2} - \ldots - \theta_q a_{t-q}$ is a *moving average process of order q, (MA(q))*. A model that includes terms of both types $\tilde{z}_t + = \phi_1 \tilde{z}_{t-1} + \ldots + _p \tilde{z}_{t-p} + a_t - \theta_1 a_{t-1} - \ldots - \theta_q a_{t-q}$ is a *mixed autoregressive moving average model of order (p,q), (ARMA(p,q))*. The above models are used with stationary processes. Nonstationary processes can often be made stationary by defining a new series of differences $x_t = z_t - z_{t-1}$. The differencing may have to be done more than once to achieve stationarity. If the d-th difference of a series is an *ARMA(p,q)* model, the series is an *autoregressive integrated moving average model of order p,d,q (ARIMA(p,d,q))*.

It can be shown that the autocorrelation function for an *AR(p)* process is related to the parameters of the model as follows: $\rho_k = \phi_1 \rho_{k-1} + \ldots + \phi_p \rho_{k-p}$. The system of equations obtained by substituting $k = 1, 2, \ldots p$ into that equation is known as the *Yule-Walker Equations*. When r_k is substituted for ρ_k and the equations are solved for the ϕ_i, we have the *Yule-Walker estimates*, ϕ_i. The k-th coefficient in an *AR(k)* process is called the *partial autocorrelation*, and plotting it versus the lag k yields the *partial autocorrelation function*. For an *MA(q)* process, the relationship is a nonlinear one: $\rho_k = (-\theta_k + \theta_1 \theta_{k+1} + \ldots + \theta_{qk} \theta_q)/(1 + \theta_1^2 + \ldots + \theta_q^2)$ for $k = 1, 2, \ldots, q$ and zero for $k > q$. The autocorrelation function of an *AR(p)* process tails off while the partial autocorrelation function is zero (has a *cut-off*) after lag p. The *MA(q)* process exhibits just the opposite behavior: The autocorrelation function has a cut-off after lag q, and the partial autocorrelation function tails off. If both functions tail off, a mixture is indicated. Those facts, plus certain limits on the parameters and autocorrelations, give a means of *model identification*. The *minimum mean square error forecast* of the series for leadtime 1 is made by replacing t by $t + 1$ in the model and then taking the conditional expectation of z_{t+1} given a knowledge of the z_t up to time t. There are, of course, other forms of forecasting.

Another prominent scheme for modeling a time series, used by the U.S. Census Bureau, particularly for economic series, rests on the idea that the series is a mixture (or sum) of 4 components: (1) a long-term smooth trend, (2) some fluctuations about the trend, (3) a seasonal component, and (4) a random or residual component. In order to separate the components, one must first estimate the trend and then remove it. Trends are sometimes fitted to low-order polynomials, but more frequently some type of moving average that smooths out the noise is used. The trend is then subtracted from the data to give an estimate of the seasonal plus random components. (If the series is thought to be multiplicative in nature, the logarithm of the series follows the above model and division is used at that point instead of subtraction.) Another smoothing process (a moving average with fewer terms than that used to estimate trend) is used to estimate the seasonal effect, and it is subtracted (or divided) to get the random or residual effect. Harmonic analysis is also used to estimate the seasonal effect. Slutzky suggested that some fictitious periodic behavior in economic time series could arise from the smoothing procedures used on the data. That is called the *Slutzky-Yule effect*.

A final, but important, approach to time series analysis is *spectral analysis*. That approach is most useful when there are cyclic or periodic components of the signal present. A function $g(t)$ is *periodic* if $g(t) = g(t+T)$ for all t; T is then said to be the *period*, and $1/T$ is the *frequency* in cycles per unit time. If the record length is chosen to be T (the period), then $1/T$ is the *fundamental frequency* and integer multiples of $1/T$ are *harmonics* of the fundamental frequency. If the record consists of N points, equally spaced at intervals of Δ, the highest frequency that can be detected is $f = 1/2\Delta$, called the *Nyquist frequency*. As a result, certain higher frequencies are indistinguishable from lower ones with the same period and are said to be *aliases* of the lower frequencies.

For a function such as $A \sin \alpha t$, A is the *amplitude* and α is the *angular frequency* (as opposed to the "cycle" frequency mentioned above) and measured in radians per unit time. The period or *wavelength* is $2\pi/\alpha$. For the function $A \sin \alpha t + B \sin \alpha t = \rho\cos(\alpha t - \theta)$, the amplitude is $(A^2 + B^2)^{1/2}$ and θ is called the *phase*.

For a stationary stochastic process $X(t)$, the *spectrum* (or *power spectrum*) $\Gamma(f)$ is the Fourier transform of the autocovariance function $\gamma(u)$, so that $\Gamma(f) = \int_{-\infty}^{\infty}\gamma(u)e^{-i2\pi fu}du$. The spectrum shows how the variance (or average ac power) of the process is distributed over frequency f. The same Fourier-pair relationship holds for the sample autocovariance function $c(u)$ and the sample spectrum $C(u)$. Smooth series have spectra with most of the power at low frequencies while highly oscillating series would show most of their power at high frequencies. The sample variance or *average power* of a stationary time series $x(t)$ in the interval $-T/2 \leqslant t \leqslant T/2$ can be represented as the sum of contributions from the harmonics $f_k = k/T$ as follows: $s_T^2 = $

$$T^{-1} \int_{-T/2}^{T/2} x^2(t)dt = \sum_{-\infty}^{\infty} |X_k|^2,$$ where X_k is the *complex amplitude* and $X_k = $

$$T^{-1}\int_{-T/2}^{T/2} x(t) \ e^{-i2\pi kt/T} \ dt.$$ For a discrete signal, $s_T^2 = N^{-1} \sum_{-n}^{n-1} x_t^2 = $

$$\sum_{-n}^{n-1} |X_m|^2,$$ where $X_m = N^{-1}\sum x_t e^{-i2\pi kt/N}.$

The contribution $|X_k|^2$ to the average power for frequency f_k is the *intensity* at f_k and a plot of $|X_k|^2$ versus k is a *Fourier line spectrum*. Frequently a sum of sines and cosines is fitted to the N data points. If a_i is the coefficient of the i-th cosine term and b_i the coefficient of the i-th sine term, the *intensity* of frequency f_i is $I(f_i) = (N/2)(a_i^2 + b_i^2)$, and the plot of $I(f_i)$ versus f_i is the *periodogram*.

For a record of infinite length, the variance τ^2 is the limit, as $T \to \infty$ of s_T^2, i.e., $\tau^2 = \int_{-\infty}^{\infty} \Gamma(f)df$, where $\Gamma(f)$ is the power spectrum or spectrum. We have seen previously that a stationary stochastic process is simply described by the autocovariance function. The power spectrum is an equivalent description. The function $C(f) = T|X_m|^2$ is called the *sample spectrum*.

For records of finite length, the spectrum is usually preferred to the autocovariance function. The power spectrum divided by the variance of the process is the *spectral density function* and is the Fourier transform of the autocorrelation function.

A plot of the sample spectrum shows how the variance or average power is distributed over frequency. Large values of the spectrum indicate periodicities at that frequency. Various types of processes can be identified, to some extent, by their spectra.

In a time series x_t there is frequently a smooth trend overlaid with random error, and that leads to the idea of *smoothing* the series in order to estimate the trend and possibly remove it. Two techniques for doing so are (1) fit a smooth function, such as a low-order polynomial, to the data and (2) substitute for each data point a weighted average of points nearby, giving a high weight to the data point itself. That is a *moving average* $\sum_{-n}^{n} c_i x_{t+i}$ smoothing procedure.

The sample spectrum has a large variance, which can be reduced by smoothing. In practice, infinite-length records are reduced to finite length $(-T/2, T/2)$ records by passing the signal through a *data window*, which is zero everywhere except on the interval $(-T/2, T/2)$, where it is 1. The expected value of the sample spectrum $C(f)$ happens to be the Fourier transform of the product of the autocovariance function $\gamma(u)$ and a *lag window* or data window $w(u) = 1 - |u|/T$. That corresponds to viewing the spectrum through a *spectral window* $W(f)$, which is the Fourier transform of $w(u)$. The simplest data window is a *boxcar* or *rectangular window*, which gives equal weights to every value in the interval $(-M,M)$. A *triangular window* is one shaped like an isosceles triangle. Four other well-known windows are the *Tukey, Parzen, Hanning,* and *Hamming windows*, all of which give less weight to values further away from the data point being smoothed. Tukey's window is a cosine function. The width of the window is its *bandwidth*. If the difference between the smoothed spectrum and the true spectrum is as small as possible for all frequencies, we say that the smooth spectrum reproduces the true one with high *fidelity*. If the variance of the smoothed estimate is small, the estimator has high *stability*. While small bandwidth is desirable, it may be achieved at the price of large *sidelobes*, which allow distant frequencies of the spectra to make large contributions to the bias at f, an effect known as *leakage*.

When 2 stationary time series are considered jointly, there are also 2 *cross-covariance functions*, $\gamma_{12}(u) = E[(X_1(t) - \mu_1)(X_2(t+u) - \mu_2)]$ and $\gamma_{21}(u) = E[(X_2(t) - \mu_2)(X_1(t+u) - \mu_1)]$ of lag u, related by $\gamma_{12}(u) = \gamma_{21}(-u)$. That leads to 2 *cross correlation functions*, ρ_{12} and ρ_{21} similarly related. The *cross spectrum* $\Gamma_{12}(f)$, is the Fourier transform of $\gamma_{12}(u)$. The cross spectrum can be decomposed into a real part called the *cospectrum* and an imaginary part, the *quadrature spectrum*. The cross spectrum is also the product of the *cross amplitude spectrum* and the *phase spectrum*, where the former is the square root of the sum of squares of the cospectrum and quadrature spectrum and the latter is the arctangent of minus the quadrature spectrum divided by the cospectrum. The cross spectrum compares the amplitudes at the same frequency, and the phase spectrum tells by how much the components of one series lead or lag those of the other series at the same frequency. The *squared coherency* is the cross amplitude divided by the product of the 2 "variances" Γ_{11} and Γ_{22} and thus plays the role of a correlation coefficient. The plot of squared coherency versus frequency is the *squared coherency spectrum*. It provides a measure of the correlation between the 2 series as a function of frequency. The estimator of coherency is biased because of phase shift. *Alignment* of the 2 processes so that the largest absolute peak, at lag S, occurs at zero considerably reduces the bias.

References

A readable book that covers a wide range of approaches to time series is Kendall, M. 1976 *Time Series*. 2nd ed. New York: Hafner. The approach most often used is found in Box, G. E. P., and Jenkins, G. 1970. *Time Series Analysis*. San Francisco: Holden-Day. The spectral approach is found in Jenkins, G. W., and Watts, D. G. 1968. *Spectral Analysis and Its Applications*. San Francisco: Holden-Day. The last 2 are for the serious student.

11

Categorical Data

When the data in a sample are partitioned into categories and the number of items in each category counted, the counts are referred to as *categorical* (or *enumeration* or *attribute*) *data*. The data are classified according to 1 or more *categorical variables* and usually displayed in a rectangular array or *table*. We shall deal here with the case of 2 categorical variables, the extensions to higher dimensions being obvious. One of the variables is called a *row variable*, the other a *column variable*. A two-way table with I rows and J columns is called an $I \times J$ *contingency table*. Layouts with more than 2 variables are referred to as *multidimensional* contingency tables. In an $I \times J$ table, the ij-th *cell* is the intersection of the i-th row category and j-th column category. Within the ij-th cell of the contingency table we put the *observed cell frequency* n_{ij}. We let a dot represent summation, so that $n_{i.}$ is the sum of the observations in the i-th row and $n_{.j}$ is the sum of the n_{ij} in the j-th column of the table. The $n_{i.}$ and $n_{.j}$ are called the *marginal totals*, and $N = n_{..}$ is the sum of all the observations in the table.

Categorical data may be collected in one of several important ways. In *multinomial sampling* the total sample size N is specified before the sample is taken. After it is taken, each member of the sample is cross-classified by

113

row and column variables. In *product multinomial sampling* the row totals are specified in advance, and the sample units within a row are classified by the column variable. In *Poisson sampling* the observer samples from a Poisson process (possibly different for each cell) for a fixed length of time and counts the number of observations occurring in each cell. Curiously, all 3 sampling models lead to the same expected cell values and to the same goodness of fit statistics.

In an $I \times J$ table the question of most interest is whether the 2 classification variables are *independent*. We let p_{ij} be the probability of an observation belonging to the i-th row category and j-th column category, $p_{i.}$ the probability of belonging to the i-th row category, and $p_{.j}$ the probability of belonging to the j-th column category. The row and column variables are independent by definition if $p_{ij} = p_{i.}p_{.j}$. The null hypothesis or *hypothesis of independence* is H_0: $p_{ij} = p_{i.}p_{j}$. If the variables fail to be independent, there is some *association* between them and measuring the *degree* of association is frequently the object of the analysis. The *expected cell frequencies* are $f_{ij} = Np_{ij}$, and under the model of independence, $f_{ij} = N p_{i.}p_{.j}$. The $p_{i.}$ and $p_{.j}$ are unknown, but their maximum likelihood estimates are $\hat{p}_{i.} = n_{i.}/N$ and $\hat{p}_{.j} = n_{.j}/N$; hence the expected cell frequencies under the model of independence are $e_{ij} = n_{i.}n_{.j}/N$. The *Pearson chi-square test of independence*, $\chi^2 = \Sigma\Sigma(n_{ij} - e_{ij})^2/e_{ij}$, tests the agreement between the observed cell frequencies and the expected cell frequencies. Under the null hypothesis, χ^2 has a chi-square distribution with $(I-1)(J-1)$ degrees of freedom.

In using the χ^2 statistic, we are employing the continuous chi-square distribution as an approximation to the discrete multinomial distribution of observed cell frequencies. The *Yates Continuity Correction* seeks to improve the approximation in the chi-square test in the 2×2 case by subtracting 0.5 from $(n_{ij} - e_{ij})$ when it is positive and adding 0.5 when it is negative. When the expected frequencies are small, the approximation is not good, and *Fisher's Exact Test* for 2×2 tables is used instead. It uses the exact distribution of observed frequencies (the hypergeometric distribution if the marginals are fixed) rather than the chi-square approximation. In a 2×2 table where the data arise from *matched* samples, the cell frequencies are not independent; hence the chi-square test is not appropriate and *McNemar's Test*, $\chi^2 = (n_{11} - n_{22})^2/(n_{11} + n_{22})^2$, is used instead. In many studies a number of 2×2 tables is involved, all of which use the same variables but differ in some other respect, such as time. There are several methods for pooling or combining the information from the individual chi-square tests, among which are *Cochran's Method* and the *Mantel-Haenszel Test*, which is very similar to it. Both use the large sample normal distribution of n_{11}. The *Lancaster-Irwin Method* provides a way of partitioning an $I \times J$ table into independent 2×2 tables so that the overall chi-square statistic is partitioned into components with 1 degree of freedom each.

An alternative to the Pearson chi-square test is the *likelihood ratio statistic* $G^2 = -2 \; \Sigma\Sigma \; n_{ij}\ln(e_{ij}/n_{ij})$, which has an asymptotic chi-square distribution. Another alternative is the *Freeman-Tukey chi-square*, which is the sum of squares of the *Freeman-Tukey residuals*, the residual for the *i-j*-th cell being $z_{ij} = \sqrt{n_{ij}} + \sqrt{n_{ij} + 1} - \sqrt{4e_{ij} + 1}$; it has an asymptotic chi-square distribution.

The experimenter may be more interested in indexing the *strength* of the association between 2 categorical variables than in simply ascertaining whether there *is* a relationship from a chi-square test. Pearson's chi-square statistic for an $I \times J$ contingency table has been described above, but its magnitude depends on the total sample size N. To partly eliminate that dependence and to obtain measures with correlationlike values (ranging from zero to 1), a number of variations of the chi-square statistic χ^2 have been proposed. They ae the *mean square contingency coefficient* $\phi^2 = \chi^2/N$, the *coefficient of contingency* (or *Pearson's P*) $P = (\chi^2/N)/(1 + \chi^2/N)^{1/2}$, *Tschuprow's T* $= \phi^2/\sqrt{(I-1)(J-1)}$, *Cramer's V* $= \phi^2/\min(I-1, J-1)$, and the *phi-coefficient*, which is the square root of ϕ^2 with the sign of $n_{11}n_{22} - n_{12}n_{21}$ attached. Other measures are based upon the *estimated odds ratio* $\hat{\alpha} = n_{11}n_{22}/n_{12}n_{21}$. *Yule's Q* $= (\hat{\alpha} - 1)/(\hat{\alpha} + 1)$ and *Yule's Y* $= (\sqrt{\hat{\alpha}} - 1)/(\sqrt{\hat{\alpha}} + 1)$ are 2 such measures. Measures based on predictive ability are *Goodman and Kruskal's lambda measures*, λ_b, λ_a, and λ. The first of those, λ_b, is equal to $(\Sigma u_i - v)/(N - v)$, where u_i is the maximum observed value in the *i*-th row, and v is the maximum of the column totals. λ_a corresponds to λ_b with rows and columns interchanged. For λ, we use as a numerator the sum of the numerators of λ_a and λ_b and do likewise for the denominator of λ. Those quantities measure the extent to which a knowledge of the classification of one variable improves the ability to predict the classification of the other variable. *Gini's λ* is based on the proportion of unexplained variance. When a one-way ANOVA is applied to a contingency table, Gini's λ is the ratio of "between" sums of squares to "total" sums of squares.

A special case of association is that of agreement between 2 observers who are independently classifying items into I categories. The data can be laid out in an $I \times I$ table with the 2 observers as variables. *Cohen's K* $= (N\Sigma n_{ii} - \Sigma n_i.n_{.i})/((N^2 - \Sigma n_i.n_{.i})$ is the measurement of agreement proposed for that situation.

When the observations are paired or matched and measured on a continuous scale instead of being categorized, *Pearson's correlation coefficient* $r = \Sigma(x_i - \bar{x})(y_i - \bar{y})/(n - 1)s_x s_y$ is a measure of the linear relationship between x and y. *Spearman's rho* is the analogous nonparametric measure of correlation.

If the data in Pearson's r are replaced by ranks and if there are no ties, Spearman's rho results.

For the special case of an $I \times J$ contingency table in which the levels of the factors are ordered (e.g., low, medium, high), there are 3 widely used measures of association akin to correlation. We first define $S = P - Q$, where P is the number of concordant pairs of observations in the table and Q is the number of pairs of discordant observations. (A pair of observations $(a,b),(c,d)$ is *concordant* if $a > c$ and $b > d$ and *discordant otherwise*.) Let y_{ij} be the sum of elements below and to the right of the element x_{ij} in the table, and let $a_i = x_{ij}y_{ij}$. Then $P = \Sigma a_i$. Let z_{ij} be the sum of the elements below and to the left of x_{ij} in the table, and let $b_i = x_{ij}z_{ij}$. Then $Q = \Sigma b_i$. Let $N = x_{..}$ be the sum of the numbers in the table. Then *Kendall's* $\tau_a = 2S/(N - 1)$; *Kendall's* $\tau_b = 2S/\sqrt{(P + Q + X_0)(P + Q + Y_0)}$, where X_0 is the number of observations for which the first variable only is tied and Y_0 the number of observations for which the second variable only is tied, and *Kendall's* $\tau_c = 2m\,S/N^2(m - 1)$, where $m = \min(I,J)$. Kendall's τ_a is not suitable for contingency tables, since there can be ties. *Goodman and Kruskal's* Y is equal to $S/(P + Q)$. *Somer's* d_{yx} is $S/(P + Q + Y_0)$ in which Y is regarded as the "dependent" variable. Those measures are all regarded as descriptive statistics, their "significance" taken as unimportant.

Having discussed measures of association and tests of independence, we now take a different approach, which is intended to handle situations more complex than simple dependence. We shall *model* the data in a contingency table. For an $I \times J$ table, the condition for independence is that $p_{ij} = p_{i.}p_{.j}$. In terms of theoretical frequencies that is $f_{ij} = Np_{ij} = f_{i.}f_{.j}/N$. Taking logarithms, $\ln f_{ij} = \ln f_{i.} + \ln f_{.j} - \ln N$, which can be written as $u + u_{1(i)} + u_{2(i)}$, where u is the overall average of the log frequencies, $u + u_{1(i)}$ is the average of $\ln f_{ij}$ over the i-th row, and $u + u_{2(j)}$ the average of $\ln f_{ij}$ over the j-th column. That provides a linear model of the log frequencies called a *log-linear model*. A more complete model for the two-way layout is $\ln f_{ij} = u + u_{1(i)} + u_{2(j)} + u_{12(ij)}$. The models are very much like those used in the analysis of variance, hence the $u_{1(i)}$ and $u_{2(j)}$ are referred to as *main effects* and $u_{12(ij)}$ as an *interaction term* and u as the *overall mean*. A *saturated model* is one that includes the main effects and all possible interactions. As in the analysis of variance, $\Sigma u_{1(i)} = \Sigma u_{2(j)} = \Sigma_j u_{12(ij)} = \Sigma_i u_{12(ij)} = 0$. The models without interaction terms are the models for independence. The log linear model allows one to *model* the dependence structure when it exists rather than merely detecting its existence with a test of hypothesis. Fitting a log-linear model means that the *u-terms* in the model will be estimated from the data in the table by substituting the expected values (under the model) for the f_{ij}. A difficulty that may arise is that from an empty cell. Zero entries in contingency tables are of 2 types. *Fixed* (or *structural* or *a priori*) *zeros* occur when it is impossible for observations to occur in certain categories, so that the zero is due to an honest zero probability for that cell. *Sampling* (or *random*) *zeros* are due to sampling variation associated with a low probability for a given cell. They disappear when the sample size is increased sufficiently. When a contingency table contains fixed zeros, it is said to be *incomplete*. Empty cells are usually dealt with by the addition of a small quantity to all entries.

In an $I \times J$ table, 2 or more categories are *collapsible* (may be combined into 1 category) if the rows and columns are independent, but they *may* be collapsible otherwise. A family of loglinear models is said to be a family of *hierarchical models* if setting any u-term equal to zero requires that its higher order *relatives* (terms that include its subscripts) be set equal to zero, and if a u-term is not zero, its lower order relatives must be present in the model. In a 3-way hierarchical model, u_{12} is a higher order *relative* of u_1 and u_2, since the subscript 1,2 includes both 1 and 2. Consequently, u_1 and u_2 are lower order relatives of u_{12}. If we decide to leave out the u_1 term, we must also omit the u_{12} and u_{13} and u_{123} term; if we keep the u_{12} term in the model, we must retain the u_1 and u_2 terms. That is referred to as the *hierarchy principle*. For some terms in the model there are direct estimates available; for others, maximum likelihood estimates of the cell frequencies are obtained by equating expected and observed marginals, a procedure known as *iterative proportional fitting*.

An interesting algorithm for estimation when the data are incomplete or when there are missing values is the Dempster *EM algorithm*. The missing data must be missing at random, not because of the values that would have been observed (censoring). In the multinomial model data may be missing from a category because all we have is an aggregate of numbers for 2 cells rather than individual cell numbers. In the normal model the problem is the accidental or unintended loss of data; the design would be balanced otherwise. The procedure consists of 2 steps: the *E-step* (for expectation) and the *M-step* (for maximization), hence the name *EM*. For multinomial data the *E-*step consists of getting the expected values of the 2 cells given their total and the model for the complete data that involves the parameters or cell probabilities. The *M*-step uses those estimates of the complete data and estimates the parameters (cell probabilities) by maximum likelihood. Those parameter estimates can be used in turn to reestimate the expected values. That iterative process continues until it converges. In the normal model we fill in the missing cells with their expectations, using the present values of the parameters (*E*-step) and then estimate the parameters by least squares (*M*-step), continuing until convergence is reached. The method can also be used with the multivariate normal and grouped or censored data.

A rather different usage of contingency tables is made with *latent structure analysis*. A *latent variable* is one that falls into only 1 or 2 classes (such as "liberal" or "conservative") but that is not directly observable. There is no single standard, no yardstick, no perfect set of criteria by which to measure whether a man is liberal or conservative. Within a class the subjects are assumed to be independent. In order to classify persons into 1 of the 2 categories, several (more than 2) test items that are observable and require dichotomous responses are used. They make up the *manifest data*. An example is how a person voted on a certain issue. The *accounting equations* relate the latent probabilities to the manifest probabilities. For each of the possible response patterns a *recruitment probability* that is the probability of belonging to each of the latent classes can then be computed.

References

A first book might be Everitt, B. S. 1977. *The Analysis of Contingency Tables*. London: Chapman and Hall. A second book, specializing in success failure data is Cox, D. R. 1970. *Analysis of Binary Data*. London: Chapman and Hall. A good reference to log-linear models is Feinberg, S. E. 1980. *The Analysis of Cross-Classified Data*, 2nd ed. Cambridge, Mass.: MIT Press.

12

Epidemiology

Epidemiology is the study of disease and health in human populations. The word "health" means something positive in addition to the mere absence of disease. The purpose of epidemiologic research is (1) to describe the health and disease status of a population, (2) to determine the *etiology* (causes) of diseases, (3) to predict the occurrence of disease, and (4) to attempt to control diseases through understanding and intervention. Much of epidemiology is, of necessity, observational rather than experimental.

For some time, *Koch's postulates*, formulated in 1877, dominated the thinking about causation of disease, but they were concerned mostly with transmissible diseases. They stated that a *cause* or agent of a disease must satisfy the following: (1) the agent must be found in every case of the disease by isolation in pure culture, (2) the agent must not be found in cases of other diseases, (3) after isolation the agent must be able to reproduce the disease, and (4) the agent must be found in experimental subjects after the disease is produced. A statement of the beliefs of modern epidemiologists is summarized in *Evan's postulates*, which are a rephrasing of Koch's posulates in epidemiologic language. In terms to be defined shortly, they state that: (1) in a cross-sectional study, prevalence of the disease should be higher among the exposed than among the unexposed, (2) in a case-control study exposure to the assumed cause should be more frequent among the cases than among the controls, and (3) in a follow-up study the incidence rate among the exposed should be higher than among the unexposed. Following exposure, (4) the incubation periods should have a bell-shaped frequency distribution, (5) the spectrum of disease response will vary from mild to severe, (6) host responses that signal or follow the disease (e.g., antibodies) will be increased among the exposed and (7) experimental reproduction of the disease will occur more often among the exposed than the unexposed. By way of intervention (8) eliminating or modifying the cause or exposure should lower the incidence, and (9) modifying the response in the body of the host (e.g., immunization) should decrease or eliminate the effects of disease. Finally, (10) the findings should make sense both biologically and epidemiologically.

Current thinking, influenced more by cancer with its long latency period, is that more than 1 factor may be necessary for inducing a disease. It may be that factors *A, B,* and *C,* acting in that order, will produce disease *X,* and factors *A, B,* and *D,* in that order, will also produce disease *X.* For example, smoking is *a* cause of lung cancer, but not *the* cause, because not all people who smoke get lung cancer and some people who have never smoked get lung cancer. A factor is a *necessary cause* of a disease if every occurrence of the disease is preceded by that factor. Likewise, a factor is a *sufficient cause* for a disease if the presence of the factor inevitably leads to the disease. With many diseases there seem to be *clusters* or *constellations* of *factors,* acting in concert, which cause the disease. A near prerequisite to demonstrating a cause-effect relationship is that there be a *dose-response relationship* that shows increasing response with increasing dose. Ofttimes, however, there is no effect at all until a certain dose, called a *threshold,* is exceeded. A factor occurring in some but not all of the sufficient clusters is a *contributory cause*. Smoking is a contributory cause of lung cancer but is neither a necessary nor a sufficient cause. Factors involved with a disease (age, sex, marital status, etc.) are said to be *predisposing factors* if they tend to make the host more prone to get the disease or more apt to have a severe case of the disease. Factors such as climate and nutrition that facilitate the manifestation of the disease or recovery therefrom are called *enabling factors. Precipitating factors* are those associated with the onset of disease (exposure, trauma, etc.). *Reinforcing factors* (repeated exposure, etc.) aggravate the disease or perpetuate it.

A number of authors define the *latent period* to be the time from exposure to manifestation of the disease. We prefer to break that period into 2 parts: (1) the *induction period* (from the action of the cause to the initiation of the disease process) and (2) the *latent period* (from the initiation of the disease until its manifestation); we then call the total time from exposure to manifestation the *empirical induction period.* For infectious diseases there is an *incubation period* between the introduction of the germs and the appearance of the symptoms. There is thus a latent period but no induction period.

Because epidemiology is concerned with groups of human subjects, a number of terms have evolved for describing various aspects of human populations. A population being followed is called a *cohort* (the word can also mean that part of a population born during a specified time period, called a *birth cohort*). The cohort is *fixed* if all the subjects in the study enter at the same time; it is *dynamic* otherwise. In either case the subjects leave the study at different times through dying, getting the disease, arriving at the end of the study or becoming *lost-to-follow-up* (disease status not determinable). Within the time period studied, the amount of disease-free time observed for an individual is his *person-time at risk* (in a mortality study the disease is death, and the corresponding quantity is person years of life). The sum, over all the individuals in the study, of person-times at risk is the *person-time*, usually expressed as person-years, person-months, passenger-miles, etc.

There are several measures of disease *frequency*. For a specified time period, the *risk* for an individual is the probability that he will develop the disease or die within the given time period. That probability is conditional on the individual's (1) being disease free at the beginning of the period and (2) not dying of other causes during the time period. The *risk odds* is the ratio of risk to 1 minus risk. There are at least 2 very different ways of estimating the risk. The first method is to use the proportion of persons who develop the disease during the given time period, i.e., the number of new cases (incidence) during the period divided by the number of persons initially at risk. That measure is called the *cumulative incidence* (CI) or *cumulative incidence rate*, although it is not a rate. In mathematical terms a *rate* is an instantaneous amount of change in a quantity per unit change in time. Velocity, for example, is the rate of change in distance per unit change in time. A rate refers to a point in time. Risk may also be estimated by *incidence rate*, the rate of change in disease status per unit of time; it is also called *hazard rate, force of morbidity,* or *force of mortality* (if death is the disease of interest). Instantaneous rates have to be evaluated in practice by an *average rate*. The *incidence density* (ID) is the number of new cases developing in a given period of time divided by the person-time accumulated over the same interval. Both CI and ID have been called incidence rate, but $0 \leq CI \leq 1$, while $0 \leq ID \leq \infty$ since ID is expressed in units of (1/time). The 2 measures are related as follows: $CI = 1 - \exp(\int_0^t ID \, dt)$.

For census-type data, person-time is estimated as the average population in the time period multiplied by the length of the time period. Average population may be the population at the midpoint of the interval or the average of the endpoints of the interval.

Although *prevalence* is sometimes defined as the number of existing cases of the disease at time t, the definition seems pointless. For example, if the prevalence is 6 cases, one immediately asks, "Per what?" More commonly, the *prevalence* or *point prevalence* is the probability that an individual will be a case at time t; it is estimated as the proportion of the population affected by the disease at time t, i.e., the number of existing cases at time t divided by the number of persons at risk at time t. A less-used measure is *period prevalence*, the probability that an individual will be a case during a given time period; the numerator is the number of persons with the disease during the time period. *Lifetime prevalence* is the probability at time t that an individual has ever been a case; the numerator is the number of persons who have ever had the disease. The *prevalance odds* is prevalence divided by 1 minus prevalence. The *prevalence difference* is the prevalence in the exposed population minus the prevalence in the unexposed population. The *prevalence ratio* is the ratio of prevalence in the exposed group to that in the unexposed group. In a steady state setting (no changes with time) prevalence P is related to incidence ID through the mean duration of the illness $\overline{T}$: $P = (ID)\overline{T}/[(ID)\overline{T} + 1]$.

Where part of a population is exposed to a *risk factor* (a factor postulated to cause the disease) the ratio of incidence density in the exposed group to that in the unexposed group is the *incidence density ratio*. The ratio of cumulative incidence rate among the exposed to that among the unexposed is often referred to as the *relative risk*. With *no* time changes in the rates, the limit (as the length of the time period approaches zero) of the relative risk is the incidence density ratio. Both ratios have been called the *incidence rate ratio, rate-ratio*, or *risk ratio*, and the meaning has to be taken from the context.

Let p be the probability of an event occurring and $q = 1 - p$ the probability that it will not occur. Then p/q is the *odds* that it will happen. If p_1 is the incidence rate among the exposed and p_2 the incidence rate among the unexposed, then p_1/q_1 is the odds in the exposed and p_2/q_2 the odds in the unexposed and their ratio, $p_1 q_2 / p_2 q_1$ is the *odds ratio* or *relative odds*. For rare diseases the odds ratio is a close approximation to relative risk. It is important because it can be estimated from either a case-control study or a cohort study. In a case-control study with incident cases the odds ratio is also called the *exposure odds ratio* (EOR) and estimates incidence density ratio. In a fixed-cohort study the odds ratio estimates the *risk odds ratio* ROR $= [R_1/(1 - R_1)]/[R_0/(1 - R_0)]$, where R_1 is the risk among the exposed and R_0 the risk among the unexposed. In a cross-sectional study the *prevalence odds ratio* (POR) is the odds ratio at a fixed point in time and can be used, with mean duration, to estimate the incidence density ratio.

An *absolute effect* is the difference between the actual rate (among the exposed) and the null rate (among the unexposed). Measures of absolute effect are the *average rate difference* (exposed minus unexposed) and *risk difference*. Those are estimated respectively by the *incidence density difference* ($ID_1 - ID_0$) and *cumulative incidence difference* ($CI_1 - CI_0$), where the subscript 1 refers to the exposed group and 0 to the unexposed group. Similarly, there is the *prevalence difference* ($P_1 - P_0$). $CI_1 - CI_0$ has been called *attributable risk*, as have other measures. The *relative effect*, relative to the unexposed, is ($ID_1 - ID_0$)/ID_0 or ($CI_1 - CI_0$)/CI_0. The *etiologic fraction* (or *attributable risk* or *population attributable risk* or *attributable fraction*) for a dynamic population is the proportion of new cases in the time period considered that is due to or attributable to the risk factor of concern. It is estimated as ($\tilde{ID} - \tilde{ID}_0$)/$\tilde{ID}$, where $\tilde{ID}$ is the estimate of the incidence density in the whole population and $\tilde{ID}_0$ is the estimate in the unexposed population. If the disease rate is larger among the unexposed group than the exposed group, the *prevented fraction* is the proportion of potential new cases that were prevented from occurring by the exposure. It is estimated as ($\tilde{ID}_0 - \tilde{ID}$)/$\tilde{ID}_0$. The *preventable fraction* is the fraction of the diseased that would have been prevented if they had been exposed to a factor believed to be protective against the disease. It is estimated by ($\tilde{ID} - \tilde{ID}_1$)/$\tilde{ID}$, where $\tilde{ID}_1$ is the incidence rate among those exposed to the factor.

Specific rates are distinguished from *crude rates* (or *overall* or *total rates*) by being defined for particular subsets or categories of the population. Examples are *age-specific rates* (confined to age groups) or *sex-specific rates*. Crude rates are defined for the entire population, so that a crude rate is a weighted average of age-specific rates, the weights being the fraction of the total population in the category. A *directly standardized rate* is a weighted average of the specific rates, the weights being the fraction of some *standard population* (such as the U.S. population) that lies in the category of interest. The weights add to 1. The standardized rate is thus a hypothetical crude rate that would have resulted had the population at hand been a miniature of the standard population with regard to the fraction in each category. An *indirectly standardized rate* is again a weighted average in which the standard population furnishes the category specific rates and the study population furnishes the weights. A familiar example of the latter is the *standardized mortality ratio* (SMR) in which the U.S. population furnishes age, sex, race, and calendar-year specific rates while the study population furnishes person-years for each category as weights w_i. Then $\Sigma w_i r_i$ is the "expected" number of deaths (the denominator of the SMR) while the observed number of deaths in the study population is the numerator of the SMR. The SMR is thus the ratio of observed to expected deaths if the population under study had the rates that the standard population has.

The *mortality rate* or *force of mortality* is the force of morbidity in which a new "case" is a death. That measure may refer to deaths that are *due to* the disease or to deaths of persons who *have* the disease. The *fatality rate* is the force of mortality among those with the disease. The *total death rate* is the force of mortality for all diseases among the total population. The *mortality density due to the disease* is the incidence density where cases are deaths resulting from the disease. There is also a *mortality density with the disease* that is similarly defined. The *fatality density due to the disease* is an incidence density in which new cases are deaths resulting from the disease and person-time is the follow-up experience of the new cases only. The *average death rate* (or *density*) resulting from all diseases is an incidence density in which cases are deaths from any cause during the follow-up period.

There are many ways of doing epidemiologic research, each adapted to a different purpose. A *clinical trial* is used primarily to evaluate 1 or more treatments for a disease. A total sample of N persons with the disease is selected and divided at random into groups that are to receive the several treatments, 1 of which usually serves as a control. The clinical trial is *single blind* if the receiver of the treatment does not know the identity of the treatment and *double blind* if neither the subject nor the administrator (or evaluator) of the treatment knows which treatment is being given. A *field trial* is used to evaluate preventive measures for diseases that are very common or very serious in nature. Persons who do not have the disease are used as subjects. Assignments of subjects to treatments may be random, but assignments are sometimes made for convenience. A *community intervention trial* is used to study the effects of a treatment administered to an entire community (e.g., water fluoridation). A similar community is frequently used as a control.

Most other types of epidemiological studies are observational in nature rather than experimental. They involve "natural experiments" in which the investigator does not decide who is assigned to the treatments. Sometimes the subjects assign themselves (e.g., by their diet or smoking practices), and in other cases they are assigned by circumstances (e.g., radiation exposure).

The *follow-up* or *cohort study* involves subjects initially free of the disease, some of whom have been exposed, in varying degree, to a risk factor. The subjects are then followed for a period of time to obtain rates among the exposed and unexposed and compare them. In many cases, to save expenses, the "unexposed" group is taken to be the U.S. population or a state population. Those populations, by virtue of their size, also offer more stable rates.

The most efficient and most widely used method of epidemiologic research is the *case-control study*, also known as the *case-history, case-referent*, or *retrospective study*. The follow-up or cohort study proceeds from postulated cause to effect, whereas the case-control study works from effects to causes. The case-control study starts with the identification of the *cases* or subjects with the disease of interest. The cases are compared with a group of *controls* who do not have the disease. The two groups are compared with respect to exposure to 1 or more factors that are possible causes of the causes. A case-control study designed to investigate a specific hypothesis is said to be an *analytic study* as opposed to a "fishing expedition" or *exploratory study* in which the investigator has no information that would lead him to propose a cause in advance; he is merely combing the data for clues.

In a case-control study, the controls may be selected from a frame by random sampling, systematic sampling, stratified sampling, or matching. *Matching* is the use of constraints to select controls so that the *index series* (cases) are similar in some way to the *referent series* (controls). In a cohort study the index series is the exposed group. In an experiment it is the treated group. It involves pairing 1 or more controls to each case with respect to risk factors that are independent of exposure (such as age, race, sex, blood group, income, and occupational factors). If there is 1 case per control, we have *1-to-1 matching*; r controls per case results in *r-to-1-matching*. If the ratio of controls to cases is the same for every case, there is a *fixed matching ratio*; otherwise we have a *variable matching ratio*. With *stratified sampling*, the controls are selected at random from subgroups or *strata* defined in advance. The target population is stratified on the basis of a risk factor such as those used in matching, but the controls are not paired to any single case. Different numbers of controls may be selected from each stratum. The objective of matching and stratified sampling is the control of *confounding*.

A third type of study design is the *cross-sectional study* or *survey* or *prevalence study*. At a fixed point in time, a random sample is taken from a dynamic population and classified with respect to current diseases status *and* level of exposure. No follow-up period is involved.

The *directionality* of a study is also used for classification purposes. The study is *forward* in time (a follow-up study) if risk factors are observed before diseases status is known, *backward* in time (case-control) if disease status is known before the risk factors are, or a combination of the two (case-control nested in a follow-up study), which is said to be *ambidirectional*. If risk factors and disease status are observed simultaneously, the study is *nondirectional*. The *design* of a study is documented in its *protocol*. Technically, a *retrospective study* is based on records made before the protocol was written and can have any of the 4 directionality patterns. A *prospective study* is one in which the protocol precedes the recording of the data; the observations take place during the study period. Studies with both features are *ambispective*. Historically *retrospective* and *prospective* have also been used as synonyms of case-control and follow-up studies. When observations on the subjects are made at several points in time, we have a *longitudinal study*. Such a study is used to detect changes or observe growth in the subjects.

A *proportional morbidity study* involves observations on diseased cases while a *proportional mortality study* uses mortality data. The measure used is the ratio of cases to the total number of illnesses or deaths in the study population; it is compared with the corresponding ratio in a standard population.

A *screening test* is a rapid diagnostic tool applied to all subjects to categorize them as *"probably having"* or *"probably not having"* the disease. For a screening test, let A be the event of the person giving a positive response and $\overline{A}$ the event of his giving a negative response. Let B be the event of the person having the disease and $\overline{B}$ the event of not having it. The *sensitivity* of the test is the probability of a positive response when the person has the disease, i.e., $P(A|B)$. The *specificity* of the test is $P(\overline{A}|\overline{B})$, the probability of a negative response when the person does not have the disease. The *false positive rate* $P(A|\overline{B})$ is the proability of a positive response from people free of the disease, and $P(\overline{A}|B)$ is the probability of a negative response from those who have the disease, called the *false negative rate*. The *predictive valve positive* is the probability that a person has the disease given that he has a positive test, $P(B|A)$; the *predictive value negative* is the probability that the person does not have the disease given that his test results are negative, $P(\overline{B}|\overline{A})$.

A major concern of epidemiological studies is the question of various types of bias, all coming under the general heading of the *validity* of the study. There is a *target population*, about which we would *like* to make inferences, and a *study population*, from which we have observations. There is a *target parameter* θ, which we *intend* to estimate with $\hat{\theta}$. We may not be estimating θ, however, but another parameter θ_0. The estimator $\hat{\theta}$ is said to be a *valid estimator* only if $\theta_0 = \theta$, and the *relative bias* is θ is $(\theta_0 - \theta)/\theta$. A *selection bias* results from the way the subjects are selected for study; one example occurs when exposed subjects who develop the disease are less likely to be lost to follow-up than those exposed who do not develop the disease. Another example that usually shows up in occupational studies is the *healthy worker effect* in which the employed workers demonstrate lower than expected death rates. *Information bias* results when the informants give systematically inaccurate information on disease or exposure status. Other manifestations of bias are *detection bias*, which might lead one hospital staff to give more tests for brain cancer symptoms than another (that could also be called *interviewer bias*; it leads one interviewer to different results because of a subconscious diligence in searching for some types of information); *recall bias*, which would lead mothers of children with leukemia to recall better than other mothers such things a diagnostic x-rays; and *reporting bias*, which makes a subject reluctant to report his history of such things as venereal disease. A *confounder* is a risk factor that, if controlled, reduces or corrects a bias. The presence of a confounder is seen when the crude effect (such as an odds ratio) is quite different from an "adjusted" effect (the odds ratios after stratification). Confounding is frequently controlled at the design stage by stratifying on the suspected confounder. The nonuniformity of stratum-specific effects is one form of *interaction*. If the interaction involves a risk factor and if the interaction carries over to the population of interest, the population interaction is *effect modification* and the risk factor is an *effect modifier*. On the other hand, if the strata effects seem to be uniform within strata, an overall measure of effect (usually a weighted average) is calculated and a *Mantel-Haenszel Test* is used as a simmary chi-square test of overall significance of the effects.

References

An easy first book is Mausner, J. S., and Bahn, A. K. 1974. *Epidemiology, An Introductory Text*. Philadelphia: W. D. Saunders Co. A widely used beginning text is Lillienfeld, A. M., and Lillienfeld, E. L. 1980. *Foundations of Epidemiology*. Oxford, England: Oxford University Press. A readable text with the statistical details of survival analysis is Lee, E. T. 1980. *Statistical Methods for Survival Data Analysis*. Belmont, Calif.: Lifetime Learning Publications.

13

Quality Control and Acceptance Sampling

Consider a mass production line for a single "part" and let us center our interest on a single "dimension" of the part. There will always be some variability in the dimension because of the sum of several variables whose causes are not usually understood and whose effects might not be controllable if they were. The dimension of the part is thus a random variable, but one whose mean and variance can change with time as something goes wrong with the "machinery" producing the part. If the process is stable and only the "usual" system of chance causes is operating (which implies that the mean θ and variance σ^2 of the *process* are constant), we say that the process is in *statistical control* and that it has an *inherent variability* σ^2. If the mean or variance wanders from those stable values, we say that an *assignable cause* is operating, meaning that it can be understood/controlled. The term *quality control* refers to control of quality by any means, but *statistical quality control* refers to the control of quality through the use of certain statistical tools to be described shortly. The best known of those tools is the *Shewhart Control Chart*, which is used to demonstrate statistical control and to detect the presence of assignable causes. There are control charts for the mean, range, standard deviation, fraction defective, and number of defects per unit. Three horizontal lines plus the data make up the control charts. The 3 lines are a *centerline*, an *upper control limit*, and a *lower control limit*. (One of the control limits may be omitted if there is no interest in it.) The centerline is an "average" of the statistic being plotted, and the control limits are at distances of 3 standard deviations (of the statistic) above and below the centerline.

The "data," whether they be means, ranges, or whatever, are plotted in order of production of the parts so that the time behavior of the process can be observed visually. The system is assumed to be stable for at least very short periods of time; hence the observations are divided into *rational subgroups* of k consecutive observations with the most common value of k being 5.

For the $\overline{X}$-*chart,* the central limit theorem is used to obtain control limits. A group of 20 to 30 consecutive values of $\overline{X}$ are calculated, using k values per average. If those points seem to be in control, they are used to calculate the limits. The average of those points, $\overline{\overline{X}}$, serves as a centerline, and the control lines are placed at $\overline{\overline{X}} \pm 3S_x^-$, where S_x^-, is the standard deviation of the $\overline{X}_i$. Future values of $\overline{X}$ (averages of k points) are plotted on that chart. A point falling outside the control limits is considered evidence that assignable causes are operating, i.e., either the mean or variance or both have changed, and a cause for the change can be found and corrected and the process brought under control. The control chart is kept as long as the process is in operation. The $\overline{X}$-chart controls the mean of the process, while an *R-chart* controls the range, and a σ-*chart* controls the standard deviation of the process. The centerline for an R-chart is the average of the ranges of the samples, and that for a σ-chart is the average of the calculated sample standard deviations. Control limits are calculated from tables. For a *p-chart*, where p is the fraction defective, p is considered to be a binomial variable, with the centerline at $\overline{p} = $ (no. of defectives in entire sample/number of items inspected) and control limits at $\overline{p} \pm 3 \, (\overline{p}(1 - \overline{p})/n)^{1/2}$. That utilizes the normal approximation to the binomial.

There is yet another control chart called a *cusum chart* (cumulative sum), which is intended to control $\overline{X}$. If we let Y_k be the mean of the k-th rational subgroup minus the centerline $\overline{X}$, the cusum chart is a sequential plot of ΣY_k (the sum from 1 to m) versus m. The process is in control as long as the plotted points do not fall outside a V-mask, which has been constructed from a piece of cardboard, say. The mask is placed over the last plotted point so that the V is horizontal and is moved to the left from point to point. The cusum chart is especially valuable for detecting small but persistent deviations that "build up".

An item is *defective* if it does not meet specifications on 1 (or more) of its m characteristics. If we count *defects* (a characteristic out of specifications), there may be as many as m defects on 1 item. A *c-chart* controls for defects rather than defectives. Letting c' be the average number of defects per unit, we can use the normal approximation to the Poisson distribution to put control limits at $c' \pm 3\sqrt{c'}$ and a centerline at c'.

It is important that processes be left alone while the points are within limits and have the appearance of being normally distributed. Other indications of something amiss in the process are unnatural patterns in the data, such as: (1) a trend, (2) too small a variance, i.e., too few points falling between $\mu + 1\sigma$ and $\mu + 2\sigma$, (3) stratification, i.e., too few points between the centerline and the $\mu + 1\sigma$ line, and (4) bias, i.e., too many consecutive points on one side of the centerline.

A major concern in manufacturing is the quality of items bought or sold. Either the buyer or the seller (at the request of the buyer) takes samples of the items, inspects them, and decides either to accept or reject a quantity of the items. That process is termed *sampling inspection* or *acceptance sampling* and there are 2 types: (1) *lot-by-lot inspection*, in which the items are grouped into batches or *lots* (the items within a lot are produced under conditions as much alike as possible [e.g., 1 machine or 1 operator] so that the source of difficulties with a lot can be traced), and (2) *continuous inspection,* in which production is continuous and lots are not naturally formed. Current inspection results determine whether the next articles are to be sampled or *screened* (100 percent inspection).

Another method of classifying inspection plans is to divide them into: (1) *attribute sampling plans*, in which the items are classified simply as defective or not defective, good or bad, or go-no-go, and (2) *variable sampling plans*, in which items are measured on a continuous scale and the measurements recorded and used. We discuss first the attribute plans.

In lot-by-lot inspection, the object is to accept the "good" lots and reject the "bad" ones. Several types of sampling plans are used to accomplish that object. We consider first lot-by-lot inspection by attributes. A *single sampling plan* consists of the following procedure: 1 sample of size n is drawn at random from the lot of N items. The lot is accepted if the number of defectives D found in the sample is less than or equal to c, an integer called the *acceptance number*, and rejected otherwise. Both n and c depend upon N and are specified by the plan. A *double sampling plan* is carried out as follows: The first sample of n_1 items is drawn at random from the lot. The lot is accepted if the number of defectives $D_1 \leq c_1$ and rejected if $D_1 \geq r_1$. If $c_1 < D_1 < r_1$, a second sample of n_2 items is drawn. We let D_2 be the number of defectives in the combined sample of $n_1 + n_2$ items. If $D_2 \leq c_2$ the lot is accepted; otherwise it is rejected. The integer numbers n_1, n_2, c_1, r_1, and c_2 are prescribed by the plan and depend upon the lot size. In *multiple sampling plans* it is possible to take as many as k samples, with sample sizes equal to n_1, n_2, . . . n_k. Let D be the number of defectives found in the combined sample of $n_1 + n_2 + \ldots$

$\overset{\cdot}{+}$ n_i units. The plan prescribes acceptance numbers $c_1, c_2, \ldots c_k$ and rejection numbers $r_1, \ldots r_{k-1}$. Proceed as in double sampling but accept the lot if $D_2 \leq c_2$, reject the lot if $D_2 \geq r_2$, and take a third sample otherwise. At the last stage accept if $D_k \leq c_k$ and reject otherwise.

Collections or systems of attribute sampling plans are usually indexed by AQL, AOQL, LTPD, or Point of Control. It is nearly impossible to attain perfect quality; hence there will be a small percentage of defective items in every lot. What percentage of defectives is tolerable? The *acceptable quality level (AQL)* is the highest percent of defectives that is acceptable as a *process average*. Selection of the AQL is a management decision and is a measure of "good quality" lots; the probability of accepting such a lot is denoted by $1 - \alpha$, where α, the probability of rejecting "good" quality lots, is called the *producer's risk* and is conventionally set at 0.05. In a similar way a decision has to be made about "poor" quality lots: those we wish to reject most of the time. The *lot tolerance percent defective (LTPD)* is the percent defective of incoming lots that will be accepted "reluctantly," i.e., with some small probability β, where β is usually set at 0.10. The probability β of accepting lots of "poor" quality is called the *consumer's risk*. The *operating characteristic (OC) curve* of a sampling plan is a graph with percent defective of the incoming material (x) plotted against the corresponding probability of acceptance of lots of that quality (y). Typically α and β are set in advance of 0.05 and 0.10. Then the AQL and LTPD are selected. Finally, the OC curve is made to pass through the 2 points (AQL, $1 - \alpha$) and (LTPD, β). The choice of the sample size n and acceptance number c is dictated by the choice of AQL and LTPD. On the other hand, one can fix n and c and the AQL and LTPD are dictated.

A question of importance is the following: What is the quality of the material that is *accepted* for use? That quality depends on the quality of the submitted lots, the sampling plan, and what is done with rejected lots. Rejected lots can be returned or *screened* (every item inspected and only good ones accepted). If they are screened, the quality (percent defective) of the accepted items is the *average outgoing quality (AOQ)*. When the AOQ is plotted versus the quality (in percent defective) of submitted lots, we have an *AOQ curve*. The highest AOQ value on that curve is called the *average outgoing quality limit (AOQL)*. The *point of control* is the lot quality that has a probability of acceptance of 0.50. The Mil-Std 105D plans are indexed by AQL, the Dodge-Romig plans by LTPD, the Phillips Standard Sampling System by point of control, the Dodge-Romig tables by AOQL (as well as LTPD).

The Mil-Std 105D attribute plans call for 3 levels of inspection: normal, tightened, or reduced. *Normal inspection* is used when the submitted quality is AQL or better. *Tightened inspection* is called for when 2 out of 5 lots have been rejected, and a return to normal inspection is made when 5 consecutive lots have been accepted. Tightened inspection lowers the probability of accepting bad lots. *Reduced inspection* is introduced when quality is substantially better than AQL. Its use is optional, and the rules for using it are complex. Reduced inspection lowers the required sample size by about 40 percent. *Mil-Std 105D* is a collection of single and double sampling plans for attributes that has been used extensively by the U.S. government.

Still another type of attribute sampling plan is an *item-by-item sequential plan*. After each item is inspected, a decision is made to accept, reject, or continue sampling. The total number of defectives observed (y) is plotted against the number of items observed (x). Two parallel lines, the *acceptance line* and the *rejection line*, are drawn on the plot. Points falling above the rejection line cause rejection of the lot, and points below the acceptance line cause the lot to be accepted. The decision to continue testing might go on forever, so a *stopping rule* is required to end the sampling.

That plan is one of a class of *sequential probability ratio tests (SPRT)* for testing a hypothesis. They are all characterized by making 1 of 3 decisions after each observation (accept, reject, or continue testing). To be specific, we have a null hypothesis that the variables being tested have pdf $f_0(x;\theta)$ versus the alternative that the pdf is $f_1(x;\theta)$. Let λ_i be the likelihood ratio after i observations have been taken, while k_0 and k_1 are fixed constants such that $0 < k_0 < k_1$. The following procedure is used: Take x_1 and get λ_1. If $\lambda_1 \le k_0$, reject H_0. If $\lambda_1 \ge k_1$ accept H_0. If $k_0 < \lambda_1 < k_1$, take x_2, compute λ_2 and follow the same rule with the same constants. Continue until H_0 is accepted or rejected. The constant k_0 is approximately $\alpha/(1 - \beta)$, and k_1 is approximately $(1 - \alpha)/\beta$. In the ordinary hypothesis tests, the sample size is fixed in advance; in the SPRT the sample size is a random variable. For very good or very bad material, the average sample size required for the same α and β will be smaller than for a fixed-sample-size test.

For all the attribute plans described above, one can calculate and plot the *average sample number (ASN)* versus percent defective of submitted lots. The ASN curve helps to see the advantage of multiple sampling plans and sequential sampling plans over single sampling plans. It generally pays (in terms of total sampling) to use double, multiple, or sequential plans if submitted quality is either very good or very bad. Single sampling plans have the advantage of simplicity.

For continuous production (such as a conveyor belt) there are *continuous sampling plans*, the best known of which are the *Dodge Continuous Sampling plans*: *CSP-1, CSP-2,* and *CSP-3*. For CSP-1, one begins by inspecting every item until i consecutive units are accepted. At that point take 1 unit at random from every x units. If a defective is found, return to 100 percent inspection. CSP-2 is similar to CSP-1 except that 2 defectives, spaced less than k units apart, are required to revert to 100 percent inspection. In CSP-3 a defective unit must be followed by inspection of the next 4 units. Those plans are indexed by AOQL and compared by the *average fraction inspected (AFI)* curve. The *Wald-Wolfiwitz Continuous Sampling Plans* and the *Girschick Continuous Sampling Plans* are other, but lesser-known systems.

The other type of sampling mentioned above is *sampling inspection by variables*. Each item in the sample is measured on a continuous scale. It is generally assumed that the measurements are normally distributed. The plans then fall into 1 of 3 categories: (1) the standard deviation σ is known, (2) the standard deviation is unknown and estimated by the sample standard deviation, and (3) the standard deviation is unknown and estimated as $\bar{R}/c$. where $\bar{R}$ is the average range of a number of subgroups and c is a constant depending on sample size. Either the upper specification limit U or the lower limit L or both are given. Using the normal distribution, one can estimate the percent defective as the fraction of the area of the normal falling outside the specification limits. For the unknown standard deviation cases, charts are used to get percent defective. For known standard deviation, normal tables are used. A *variables sampling plan* consists of the sample size n, an estimate of the percent defective p, and a maximum allowable percent defective M. If $p \leq M$, the lot is accepted. *Mil-Std-414* is a catalog of variables sampling plans indexed by AQL and lot size. There are provisions for normal, tightened, and reduced inspection. The AQL in that standard is simply a nominal value of the percent defective. Variables sampling plans require a smaller sample size than do similar attribute plans.

References

A good first book is Grant, E. L., and Leavenworth, R. S. 1979. *Statistical Quality control.* 5th ed. New York: McGraw-Hill. A classic with further detail is Duncan, A. J. 1974. *Quality Control and Industrial Statistics.* 4th ed. Homewood, Ill.: Irwin. A thought-provoking, informal book on Quality Control in the U.S. today and the need for change is found in Deming, W. E. 1983. *Quality, Productivity, and Competitive Position.* Cambridge, Mass.: MIT, Center for Advanced Engineering Study.

14

Multivariate Analysis

Since many multivariate concepts are phrased in terms of matrices, a brief introduction to matrix theory may be helpful. A *matrix* is a rectangular array of *elements*. The elements may be numbers, random variables, functions, etc. The matrix A is also denoted by (a_{ij}), meaning that a_{ij} is the element in the i-th row and j-th column of A. A matrix with r rows and c columns is said to have *dimension* $r \times c$ (read "r by c") and is called an $r \times c$ matrix. The *transpose* of the matrix A, denoted by A', is the matrix whose ij-th element is a_{ji}, i.e., the rows and columns of A have been interchanged. A *vector* is a matrix with 1 row or 1 column. A *row vector* x is a matrix with 1 row. A *column vector* is a matrix with 1 column. Two matrices are *equal* if their corresponding elements are equal. A *square* matrix has the same number of rows as columns. A square matrix equal to its transpose is said to be *symmetric*. For a square matrix A, the elements a_{ii} are the *diagonal elements*: the others are *off-diagonal elements*. A square matrix whose off-diagonal elements are zero is a *diagonal matrix*. A diagonal matrix whose diagonal elements are all equal to unity is an *identity matrix*, denoted by I. The *trace* of a square matrix is the sum of its diagonal elements. A matrix of all zeros is a *null matrix* O. For 2 matrices of the same dimension, A and B, the *matrix sum* is $A + B = (a_{ij} + b_{ij})$.

The *scalar product*, cA, where c is a constant, is the matrix (ca_{ij}). The *matrix product,* AB, is defined only when the number of columns of A equals the number of rows of B. The product, $C = AB$, is the matrix $(c_{ij}) = (\sum_{k} a_{ik} b_{kj})$. The *inner product* of 2 vectors x and y, with the same number of

135

elements, is the sum of products: $\Sigma x_i y_i$. If $\mathbf{A}$ and $\mathbf{B}$ are square matrices and if $\mathbf{AB} = \mathbf{BA} = \mathbf{I}$, $\mathbf{B}$ is said to be the *inverse of A* and $\mathbf{A}$ is the inverse of $\mathbf{B}$. The inverse of $\mathbf{A}$ is denoted by $\mathbf{A}^{-1}$. If $\mathbf{A}^2 = \mathbf{A}$, $\mathbf{A}$ is *idempotent*. Each $n \times n$ matrix A has associated with it a unique number called its *determinant*, denoted by $|\mathbf{A}|$, which is the sum of all products of the form $\pm a_{1j_1} a_{2j_2} \ldots a_{nj_n}$, where there is one and only one element from each row and 1 from each column and the sign is determined by the number of permutations of the subscripts.

A matrix whose determinant is not zero is *nonsingular*. A set of vectors $z_1, z_2, \ldots, z_r$ is *linearly dependent* if there is some nontrivial linear combination of the z's that is zero and *linearly independent* otherwise. The *rank of a matrix* is the maximum number of linearly independent columns or rows of the matrix (whichever is fewer). The determinant, $|\mathbf{A} - \lambda\mathbf{I}|$, where $\mathbf{A}$ is a $p \times p$ matrix, is a p-th degree polynomial in λ called the *characteristic polynomial* of $\mathbf{A}$. The p roots of the equation $|\mathbf{A} - \lambda\mathbf{I}| = 0$ are the *characteristic roots (eigenvalues, latent roots)* of the matrix A. A characteristic root λ is a scalar such that $\mathbf{Ax} = \lambda\mathbf{x}$ for some nonnull vector $\mathbf{x}$ called a *characteristic vector* or *eigenvector* of $\mathbf{A}$. The matrix $\mathbf{A}$ is *orthogonal* if $\mathbf{A'A} = \mathbf{AA'} = \mathbf{I}$. If $\mathbf{x}$ is an $n \times 1$ vector and $\mathbf{A}$ is an $n \times n$ symmetric matrix, $\mathbf{x'Ax} = \Sigma\Sigma\, x_i x_j a_{ij}$ is a *quadratic form*. The matrix and its quadratic form are said to be *positive definite* if $\mathbf{x'Ax} > 0$ for all nonnull vectors x, and *positive semidefinite* if $\mathbf{x'Ax} \geq 0$ for all $\mathbf{x}$ and $\mathbf{x'Ax} = 0$ for some nonnull vector $\mathbf{x}$. The *derivative of a scalar quantity z with respect to a $p \times 1$ vector* $\mathbf{x}$ is a $p \times 1$ vector whose i-th element is $\partial z/\partial x_i$.

It is now necessary to consider several random variables. The *p-dimensional random variable* (or *random vector*) $\mathbf{X}$ is the row vector $\mathbf{X'} = (X_1, X_2, \ldots, X_p)$, where $X_1, \ldots, X_p$ are random variables defined on the same probability space. When $p = 2$ and 3, we speak respectively of *bivariate* and *trivariate* random variables. Given a random vector of continuous random variables, we say that the $X_1, \ldots, X_p$ are *jointly distributed*, meaning that for any constants $x_1, \ldots x_p$, $F(x_1, x_2, \ldots x_p) = \int_{-\infty}^{x_p} \ldots \int_{-\infty}^{x_1} f(u_1, \ldots u_p)\, du_1 \ldots du_p$ is the probability that $(X_1 \leq x_1$ and $X_2 \leq x_2, \ldots$, and $X_p \leq x_p)$, where $f(u_1, u_2, \ldots u_p)$ is the *joint probability density function* that must be nonnegative and integrate to 1. $F(x_1, \ldots x_p)$ is the *joint cumulative distribution function*. The joint cdf is zero if any argument is $-\infty$, 1 if all arguments are ∞, and right continuous in each argument. With multivariate distributions we also need to define marginal and conditional distributions. The *marginal density function* of a subset of $(X_1, \ldots, X_p)$ is found by integrating the joint pdf over the entire range of the random variables not in the subset. The *marginal cdf* of X_i, say, is the joint cdf evaluated at ∞ for every variable except the i-th. The X's are *independent* if and only if $F(x_i, x_2, \ldots x_p) = F(x_1)F(x_2) \ldots F(x_p)$ and $f(x_1, x_2, \ldots, x_p) = f(x_1)f(x_2) \ldots f(x_p)$.

Consider an event A such that $P(A)$ is not zero and let $P(AC)$ be the probability that both A and C occur. We define $P(C|A) = P(AC)/P(A)$ to be the *conditional probability* of C given that A has occurred. Two events, A and C, are *statistically independent* if $P(C|A) = P(C)$ or $P(AC) = P(A)P(C)$. The *conditional probability density function* of X_2 given X_1 is $f(x_2|x_1) = f(x_1,x_2)/f(x_1)$. The *conditional cumulative distribution* is found by integrating the conditional pdf from $-\infty$ to x.

The expected value of the random vector $\mathbf{X}$ is $E\mathbf{X} = (EX_1, \ldots, EX_p)'$. Likewise, the expected value of a matrix of random variables is the matrix of expected values of the random variables. The extension of the variance of a univariate random variable to that of a p-dimensional random variable $\mathbf{X}$ is the *variance-covariance matrix of* $\mathbf{X}$, denoted by $\boldsymbol{\Sigma} = (\sigma_{ij})$, where σ_{ij} is the covariance of X_i and X_j, and σ_{ii} is the variance of X_i. The *correlation matrix* of $\mathbf{X}$ is the matrix $\mathbf{R} = (\rho_{ij})$, where ρ_{ij} is the correlation between X_i and X_j. The *conditional expectation* of the function $g(X_1,X_2)$, given $X_1 = x_1$, is $\int_{-\infty}^{\infty} g(x_1,x_2)f(x_2|x_1)dx_2$. When they exist, $E(X_2|X_1 = x_1)$ is the *conditional mean* of X_2 given $X_1 = x_1$, and $E(X_2^2|X_1 = x_1) - [E(X_2|X_1 = x_1)]^2$ is the *conditional variance* of X_2 given $X_1 = x_1$.

If we let $\mathbf{X} = (X_1, X_2, \ldots X_p)$ be a vector of normal random variables, $\boldsymbol{\mu} = (\mu_1, \ldots \mu_p)$ be the vector of means, and $\boldsymbol{\Sigma}$ their covariance matrix, the function $f(X_1, X_2, \ldots X_p) = (|\boldsymbol{\Sigma}|^{1/2} (2\pi)^{p/2})^{-1} \exp [-\frac{1}{2}(\mathbf{X} - \boldsymbol{\mu})' \boldsymbol{\Sigma}^{-1}(\mathbf{X} - \boldsymbol{\mu})]$ is a *multivariate normal density function*. It can be shown that the marginal distribution of any X_i is univariate normal, but normality of the marginals does not imply multivariate normality. If the vector $\mathbf{Y}$ is multivariate normal, the components of $\mathbf{Y}$ are jointly independent if and only if the covariance of Y_i and Y_j is zero for $i \neq j$. The multivariate normal density is constant on the ellipsoids $(\mathbf{X} - \boldsymbol{\mu})'\boldsymbol{\Sigma}^{-1}(\mathbf{X} - \boldsymbol{\mu}) = c$, where c is a constant. The ellipsoids have as their center the vector $\boldsymbol{\mu}$; their shape and orientation is determined by $\boldsymbol{\Sigma}$ and their size determined by c. The ellipsoids are sometimes called *concentration ellipsoids*. Their axes are called the *principal axes of the multinormal density*. The longest axis is determined by the largest characteristic root of $\boldsymbol{\Sigma}$, the second longest by the second largest root, etc.

Let the $(p + q) \times 1$ vector $\mathbf{Y}$ have a multivariate normal distribution (with $\boldsymbol{\Sigma}$ nonsingular) and let $\mathbf{Y}$ be partitioned thus: $\mathbf{Y}' = (\mathbf{Y}'_1, \mathbf{Y}'_2)$, where $\mathbf{Y}_1$ has p components and $\mathbf{Y}_2$ has the remaining q components. The distribution of $\mathbf{Y}_1$ given $\mathbf{Y}_2 = y_2$ is multinormal, and the elements of its covariance matrix are called *partial variances and covariances*; the covariance between the i-th and j-th variate in $\mathbf{Y}_1$, given that the variates in $\mathbf{Y}_2$ are held constant, is denoted by $\sigma_{ij \cdot p+1, \ldots, p+q}$. The corresponding *partial correlation coefficient* is defined as the partial covariance divided by the square root of the product of the partial variances. Consider the linear combination of variates in $\mathbf{Y}_2$ that maximize the correlation of $\mathbf{Y}_2$ with the i-th variate in $\mathbf{Y}_1$. The coefficients in that linear combination are called the *regression coefficients* of the i-th variate in $\mathbf{Y}_1$ upon the elements of $\mathbf{Y}_2$, and the correlation is called the *multiple correlation coefficient*.

Much of the work in multivariate analysis is devoted to extensions of univariate statistics and hypothesis tests. A multivariate test of hypothesis that the mean vector μ is equal to some specified constant vector μ_0 would ordinarily require us to make n simultaneous univariate tests. To avoid the labor involved, a single test can usually be constructed by means of the *union-intersection principle*, which uses the intersection of the univariate critical regions as the multivariate critical region, and the maximum (over all linear combinations of $\mathbf{X}$) of the square of the t-statistics, $N(\bar{\mathbf{x}} - \mu_0)'\mathbf{S}^{-1}(\bar{\mathbf{x}} - \mu_0)$, as a test statistic called *Hotellings T^2* where $\mathbf{S}$ is the sample estimate of the covariance matrix. In the univariate analysis of variance the means of several populations are compared. In the multivariate extension we test several mean vectors for equality, assuming equal covariance matrices. Using the union-intersection principle with the F-test leads to a test statistic that is the largest characteristic root of the matrix $\mathbf{HE}^{-1}$, where $\mathbf{H}$ is a matrix of sums of squares and products and $\mathbf{E}$ is the error matrix. Two other commonly used criteria for the test are (1) *Wilk's Lambda Criterion* $= |\mathbf{E}|/|\mathbf{H} + \mathbf{E}|$ and (2) the *Lawley-Hotelling Trace Statistic* $= \text{trace } \mathbf{HE}^{-1}$. The *multivariate analysis of variance* is frequently referred to as a *MANOVA*.

Apart from the usual one-way and two-way analysis of variance, there are 2 other techniques that use MANOVA. One is *profile analysis*. Suppose, for example, that we wish to compare 4 groups of subjects on the basis of their "profiles." The profile for each group is made by giving each subject a battery of k tests. The average score on each test for each group is calculated and plotted as y versus the test number x, $x = 1, 2, \ldots k$. Connecting those scores leads to a profile for that group. A profile is then constructed for each of the other groups. The analysis is begun by testing the parallelism of the profiles. If they are found to be parallel, the profiles are tested to see whether they are "identical" (could have come from the same population).

Sometimes *repeated measures* are made on one individual (he may be weighed each day). The measurements then form a *growth-curve*. Regression is not appropriate here because the observations are highly correlated, but time series methodology can be used. Frequently a MANOVA approach is used to fit the data to several linear segments because it can take into account the correlations. Many other multivariate test statistics are constructed by using that principle. The sum-of-squares-and-cross-products matrix $\mathbf{A}$ has a *Wishart Distribution*, which is the multivariate analog of the chi-square distribution.

The dependence structure of a multivariate population may be studied in various ways, of which we mention 2: principal components and factor analysis. A *principal components analysis* is a partitioning of the total variance of the system into successively smaller portions by means of an orthogonal transformation of the coordinate axes. Its purpose is to reduce the dimensionality of the system, i.e., to economize on the number of variates needed. We let $\mathbf{X} = (X_1, \ldots, X_p)$ be a multivariate vector with sample covariance matrix $\mathbf{S}$. The first *principal component* of $\mathbf{X}$ is that linear combination $Y_1 = a_1 X_1 + \ldots + a_p X_p = \mathbf{a}'\mathbf{X}$ whose sample variance is greatest among all coefficient vectors normalized so that $\mathbf{a}'\mathbf{a} = 1$. The coefficients a_i are the elements of the characteristic vector associated with the largest characteristic root λ_1 of $\mathbf{S}$. The second principal component $Y_2 = b_1 X_1 + \ldots + b_p X_p$ is chosen so that it has maximum variance among vectors b with the constraint that $\mathbf{b}'\mathbf{b} = 1$ and $\mathbf{a}'\mathbf{b} = 0$. The vector $\mathbf{b}$ is the characteristic vector of $\mathbf{S}$ associated with the second-largest characteristic root λ_2 of $\mathbf{S}$, and λ_i is the variance of Y_i.

Principal component analysis is equivalent to factoring $\mathbf{S}$ into $\mathbf{P}\ \Lambda\ \mathbf{P}'$, where Λ is the diagonal matrix with λ_i on the diagonal. An important use of that technique is to find a fewer number of variables that take up most of the total variance than the system had originally. If the last few components have very little variance associated with them, they can be dropped from future analyses. In terms of axes rotation, the principal components are the new variates specified by a rigid rotation of the original axes into an orientation that gives the most variance to the first axis. The second axis is perpendicular to the first and chosen so that it takes up the maximum amount of the remaining variance, etc.

Principal components is a transformation of the covariance matrix of the multivariate observations. A different way of treating dependence structure is to model the observations in a *factor analysis*. In the model each response variate X_i is represented as a linear combination of a few *common-factor variates* and a single latent *specific variate*. The common factors generate the covariances among the variates, and the specific term contributes only to the variances of the X_i. The model is $X_i = \Sigma \lambda_{ij} Y_j + e_i$, where the Y_j is the common factor, e_i is the specific factor, and the coefficient λ_{ij} is the covariance or the *loading* of the i-th response on the j-th common factor. The variance of the i-th response is $\sigma_i^2 = \Sigma \lambda_{ij}^2 + \psi_i$, where the $\Sigma \lambda_{ij}^2$ is the *communality* of the i-th response. The purpose of the analysis is to determine the elements λ_{ij} of the *loading matrix* λ. That in turn allows us to estimate the ψ_i. The population covariance matrix Σ is thus modeled as $\Sigma = \lambda \lambda' + \psi$. Names are then frequently devised for the common factors. The factorization is, unfortunately, not unique, and the same correlation or covariance structure can be produced by several models with the same number of common factors but different loadings or by different numbers of common factors. To obtain some degree of uniqueness, one tries to achieve *simple structure*, which results when each variable has a nonzero loading on only 1 common factor. That process, known as *rotation*, is an attempt to make each factor uniquely define a distinct cluster of intercorrelated variables. An *orthogonal rotation* results when we rotate the common factor axes in such a way that they are still perpendicular to each other. If we relax that restriction of orthogonality, we have an *oblique rotation*. If we rotate the axes by trying to maximize the variance of the square of the loadings in each column of the factor matrix, we have a *varimax* rotation. Doing a similar thing for rows gives us a *quartimax rotation*. A compromise between the two is an *equimax rotation*. Again the objective is to arrive at a few common factors that will adequately explain the structure of the correlation or covariance matrix. The tools used in that area are somewhat more subjective and less statistical than in others.

Given a random sample of N_1 and N_2 observations from 2 p-dimensional multivariate normal populations with mean vectors μ_1 and μ_2 and common covariance matrix Σ, let $\bar{x}_1$ and $\bar{x}_2$ and S be the usual estimates of μ_1, μ_2, and Σ. When faced with the problem of classifying a new observation x_0 into 1 of those 2 known populations, we use the *linear discriminant function* $y = (\bar{x}_1 - \bar{x}_2)'S^{-1}x_0$ and assign x_0 to population 1 if $y > \frac{1}{2}(\bar{x}_1 - \bar{x}_2)'S^{-1}(\bar{x}_1 + \bar{x}_2)$ and to population 2 otherwise. The discriminant function y is that linear combination of the variables that best discriminates between the 2 groups. The criteria used for y is a constant that is the midpoint of the means of the 2 samples. If we wish to classify an observation into one of k groups, there are $k - 1$ such functions that need to be evaluated. If the parameters are known *a priori*, the discriminant function is $y = (\mu_1 - \mu_2)'\Sigma^{-1}x_0$. If y is greater than $\frac{1}{2}(\mu_1 - \mu_2)'\Sigma^{-1}(\mu_1 + \mu_2)$, classify x_0 as coming from population 1. The variance of the latter discriminant function is $(\mu_1 - \mu_2)'\Sigma^{-1}(\mu_1 + \mu_2)$ and is known as the *Mahalanobis Distance* between the 2 populations. To evaluate the success of the linear discriminant function, we calculate a table of *misclassification probabilities*.

In testing the degree of independence of 2 sets of variates, a technique known as *canonical correlation* is used. If there are p variates in the first set and q variates in the second set, the vector X is partitioned into $X = (X_1, X_2)'$, where X_1 is $p \times 1$ and X_2 is $q \times 1$. The sample covariance matrix S is partitioned accordingly into $S_{11}S_{12}S'_{12}$ and S_{22}. We first construct the linear combination u_1 of variates in X_1 and the linear combination v_1 of variates in X_2 such that u_1 and v_1 have the maximum correlation of any 2 combinations. Then we construct u_2 and v_2 with maximum correlation among all the linear combinations uncorrelated with u_1 and v_1, etc., until $s = \min (p,q)$ such combinations have been constructed. The correlation between u_i and $\cdot u_j$ or between v_i and v_j is zero and that between u_i and v_i is $c_i^{1/2}$, where the c_i are the characteristic roots of $|S_{12}S_{22}^{-1}S'_{12} - \lambda S_{11}| = 0$. The $c_i^{1/2}$ are called the *canonical correlations*. In practice the coefficients of the linear combinations are not calculated, all the information being channeled into the c_i. Finally, the correlation matrix of the u_i and v_j is displayed. Very small correlations indicate that u_i and v_j are nearly independent.

Cluster analysis is the grouping of similar objects where the groups are not known in advance, as they were with discriminant analysis. A *cluster* is a set of similar objects, but the concept of "similarity" is not well defined. It is generally agreed that members of a cluster are more similar to each other than to those outside the cluster. Sometimes the algorithm defines the cluster: Whatever the algorithm groups together is a cluster. The basic data used for clustering is a set of N individuals or *cases* on each of which p variables are measured. Some or all of the variables may be dichotomous (indicate the presence or absence of a characteristic). Similarities usually take the form of distances defined for each pair, (which in some cases take on values of zero and 1 only). Each individual thus represents a "point" in space. There are numerous methods of clustering, but 2 of the more popular will illustrate: (1) given an $N \times N$ similarity matrix, such as a correlation matrix, the 2 individuals that are closest (most similar) are paired. The 2 columns and rows in which they occur are then deleted and replaced by a single column and row containing some type of "average" similarity measure. That process is then repeated on the new $(N-1) \times (N-1)$ matrix. In the second step a new pair may be clustered, or a third case added to the existing cluster. (2) In the second method the points are partitioned into 2 sets so that the sum of squares between sets is maximized. That is the same thing as minimizing the sum of squares within sets. Applications of clustering are (1) classification of animals and plants, (2) classification of diseases, and (3) classification of relics in archaeology by tribal group or civilization.

Multidimensional scaling (MDS) is a set of multivariate techniques used for parameter estimation and goodness-of-fit in a spatial model for proximity data. MDS is used to study the structure of objects (such as words) or people (such as their interpersonal relationships). A *proximity measure* must be defined for every pair of objects and measures the *dissimilarity* between them. It should be functionally related to the attributes of the objects. (If the objects are cars, their attributes may be price, gas mileage, size, etc.). Most commonly, dissimilarity is estimating by averaging the judgments of several people who rate each pair of objects on a scale. Other widely used measures of dissimilarity are distances and probabilities (joint and conditional). Given the dissimilarity matrix, an *MDS algorithm* is used to decide whether the dimensionality can be reduced (i.e., if some aspects are not salient) or whether the complexity of the relationships can be reduced (to a simpler, easier-to-understand form). MDS is related to cluster analysis but more closely to factor analysis.

Another recent and versatile idea in multivariate analysis is *projection pursuit* (PP). The idea behind PP is to be able to interpret a set of high-dimensional data (a "point cloud") through some "interesting" or "well-chosen" projection that will reduce the dimensionality and make for better understanding. That is done by maximizing some objective function called the *projection index*. In a regression context that function is the criterion of fit, which has been called *the figure of merit*. For example, if we wish to preserve, as well as we can, the distance between points, we choose the variance as a figure of merit and project it onto the largest principal component. PP is also a powerful way of "lifting" a one-dimensional technique into a higher dimension. If we start with a two-sample t-statistic, say, PP will generate a linear discriminant analysis. Cluster analysis in several dimensions can be done by looking for clusters in a projection. Oblique and quartimax factor analysis are other special cases.

Multivariate data usually present such a mass of numbers that the human mind is totally confused. A device that helps grasp a moderately large array is the use of *Chernoff Faces*. Since we are used to looking at human faces, we can detect relatively small differences in facial features. With that device, we draw a set of faces of the same size (1 for each population involved.) If we have random variables $X_1, X_2, \ldots X_k$, we let X_1 equal the length of the nose, X_2 the width of the nose, X_3 the height of the eyebrows, X_4 the length of the eyebrows, X_5 the size of the eyes, etc. Of course the variables have to be properly scaled to fit within the facial area. A person can then compare the faces (populations) and detect rather easily where the differences are. The device has been applied to rectangles, pictures of trucks (in the oil industry) etc.

References

One of the easier and better books in this field is Morrison, D. 1976. *Multivariate Statistical Methods*. 2nd ed. New York: McGraw-Hill.

15

Survey Sampling

Practically everyone is familiar with the Gallup Poll, an opinion survey, and the Current Population Survey, used by the Census Bureau to estimate unemployment, the Consumer's Price Index, etc.

Survey sampling consists of the methods used to select and observe a part of a population called a *sample*. The *sampling plan* is the set of steps taken in selecting the sample. The *sample design* is the sampling plan together with the methods used in estimating the population characteristics. The *survey design* is the sample design together with the questionnaire and method of measurement.

Most of what we learn is based on sampling. We sample the grapes before buying them; the medical technician samples our blood and makes conclusions about our health. The sample consists of individuals or objects to be measured, called *elementary units* or *elements*. The entire group of elementary units is called the *population* or *universe*. Sampling is cheaper, faster, and may even be more accurate than examining the entire population because we can afford to be more thorough in the analysis. A 100 percent sample of the population (a complete enumeration) is a *census*. The word "population" or "universe" frequently implies an infinite or conceptually infinite collection of measurements, such as the endless tossing of a fair coin. In the realm of survey sampling the population is finite (of size N), and the sample is of size n. The quantity $f = n/N$ is the *sampling fraction*. With infinite populations the variance of the sample mean is s^2/n. For finite populations that quantity has to be multiplied by a factor $1 - f = (N - n)/N$ called the *finite population correction (fpc)*; it is the proportion of the population not included in the sample. The factor also appears in normal confidence intervals and in the estimated variance of a proportion. The inverse of f, N/n, is the *expansion* (or *raising* or *inflation*) *factor*. The *target population* is what we wish or intend to sample, such as the entire U.S. population, but it may not coincide with the actual *sampled population*, which may be a sample of college students.

Once the elements are defined, the sampler needs a *frame*, or list of the elements, from which to draw a sample. Examples of available frames are telephone directories and lists of registered voters. In other cases a frame has to be constructed. There are frequently problems with frames, such as incompleteness, inclusion of foreign elements, duplicates, etc. Given a frame, there are numerous ways of drawing a sample from it. The most basic, but not the most widely used, method is to draw a *simple random sample*. In a population of size N, there are $\binom{N}{n}$ possible combinations of n elements. If we let each combination have the same chance of being selected, we end up with a *simple random sample*. In practice the sample is drawn element by element. Each of the N items in the population is numbered from 1 to N. A sample of n numbers between 1 and N is taken from a random number table (or generated by some other process). Items with those numbers constitute the sample. A simple random sample is drawn *without replacement* so that once an item is in the sample, it cannot be selected again. If an item can be reselected, we speak of a *random sample with replacement*. The latter type of sampling is also called *unrestricted sampling*, and it is then possible to have $n > N$.

If each element in the population has the same probability of being selected, we speak of *Epsem* (equal probability of selection method) *sampling*. Most of the types of sampling we describe will be epsem sampling. *Probability sampling* is sampling in which each element has a known (but not necessarily equal) probability of being selected. Probability sampling is necessary to calculate the variance of the sample mean, total, etc. Opposed to probability sampling is *judgmental or purposive sampling* in which an "expert" gives his impression of which units are to be in the sample. That technique could be used effectively, for example, in selecting 30 counties to represent the U.S. Judgmental sampling is not random; hence no statistical conclusions can be based on it. It includes the use of captive audiences, volunteers, etc.

The purpose of sampling is usually to estimate one of the following population characteristics: the mean ($\overline{Y}$), the total (Y), the variance (S^2), a proportion (p), or a ratio (r). If there are several sampling methods that could be used to estimate the same population parameter, we prefer the one with smallest variance. That criterion, however, must be balanced with cost considerations. A second common method of sampling is *stratified random sampling*. The population of size N is divided into k nonoverlapping and exhaustive subpopulations of size $N_1, N_2, \ldots N_k$. The subpopulations are called *strata*. The stratification is done on the basis of some characteristic (sex, age, race, religion, etc.) in the belief that the elements within a stratum will be more homogeneous than those across the strata. A simple random sample is then selected within each stratum. A weighted mean of the strata means $\bar{y}_w = \Sigma\, w_i \bar{y}_i$ is used instead of the overall mean $\bar{y}$ to estimate the population mean $\bar{Y}$. The weight in the i-th stratum is $w_i = N_i/N$, so that $\Sigma\, w_i = 1$. The weighted mean coincides with the overall mean only if the sampling fractions n_i/N_i for the strata are all equal to n/N, where $n = \Sigma n_i$ and $N = \Sigma N_i$. In that case we say that we have a *proportionate sample* or a *self-weighted sample*. Depending on the degree of homogeneity within the strata, the variance of the weighted mean can be considerably smaller than the variance of the overall mean. Proportionate sampling is what people usually have in mind when they speak vaguely about a *representative sample*; they desire a sample that is a miniature of the population. We discourage use of that term for two reasons: (1) every sample is representative of the population in some respect, and (2) even proportionate sampling will not lead to samples representative in every respect. Within a stratified sample disproportionate sampling is sometimes used deliberately to reduce the variance of the sample mean. A compromise between cost and precision usually has to be made. If c_i is the cost of measuring a sample unit within the i-th stratum, and c_0 an overhead cost, the total cost of sampling is $C = c_0 + \Sigma\, n_i c_i$. For stratified random sampling, the variance of the weighted mean is smallest when n_i is proportional to $N_i S_i/\sqrt{c_i}$, where S_i is the standard deviation for the i-th strata. That means we should sample more heavily the larger strata, the more variable strata, and the strata where sampling is cheaper. In particular, $n_i = (n N_i S_i/\sqrt{c_i})/(\Sigma N_i S_i/\sqrt{c_i})$. If the cost of the survey is fixed (limited), we take $n = (C - c_0)(\Sigma N_i S_i/\sqrt{c_i})/(\Sigma N_i S_i\sqrt{c_i})$ and use it in the above formula for n_i. That choice of n_i is called *optimum allocation*. If the variance V has been prescribed in advance, we choose $n = (\Sigma N_i X_i/\sqrt{c_i})(\Sigma N_i S_i/\sqrt{c_i})/(N^2 V + \Sigma N_i S_i^2)$. For the special case where the costs per unit are equal in every stratum, we have $n_i = n N_i S_i/(\Sigma N_i S_i)$, called *Neyman allocation*.

A third important method of sampling is called *systematic sampling*. The elements are numbered 1 to N. The first unit in the sample is selected at random from among the first k units. Thereafter, every k-th element is selected. More specifically, it is an *every k-th systematic sample*. That type of sample is widely used because it is so convenient and because it compares favorably in precision with the types of sampling described above. If there are periodicities in the data, that type of sampling could be a disaster. We may think of a systematic sample as an artificial stratified sample with k elements per strata. The single sample unit within each stratum is selected in a fixed instead of a random way. A mixture of the 2 methods is called *stratified systematic sampling*. The population is first stratified, and within each stratum a systematic sample is selected, using an independent random starting place within each stratum.

A fourth method of sampling is *cluster sampling*, frequently used when a frame of the elements is not available or when element sampling is expensive. Consider, for example, the problem of getting a sample of households across the United States. For simple random sampling or systematic sampling thousands of phone books would be required for a frame. With *cluster sampling* we would select a random sample of counties within the U.S. and then take *every* household within the sampled counties (*single stage sampling*); or if we take a *sample* of households within each of the sampled counties, we have *two-stage sampling*. Each county then forms a *cluster* of households. The travel time and associated costs per household would be cut dramatically. The number of units within a cluster is the *cluster size*, and they are frequently unequal.

The units sampled at the first stage (counties) are *primary sampling units* (*PSU's*), those sampled at the next stage (households) are *second-stage units*, etc. With 2 or more stages of sampling, we have *multi-stage sampling*. The *ultimate cluster* for a PSU is the aggregate of sample elements that arise from that PSU; it is also called a *primary selection* (*PS*). Multistage sampling is also called *subsampling*. In *stratified cluster sampling* the population is stratified, and clusters are selected within each stratum. In the above example states could serve as strata and counties as clusters.

Finally there is *double sampling* or *two-phase sampling*, which can be generalized to *multiphase sampling*. In the first phase a large sample is taken (sometimes the entire population), and *ancillary information* is gathered. It is necessary that sampling in the first phase be quick and cheap; it acts as a screening procedure. In the second phase a small sample is taken, using the ancillary information to improve the sampling. Suppose we wish to estimate electrical energy usage in a certain subdivision. In the first phase we could drive around in an auto or use county records to quickly decide how many houses there were in each of 3 strata. With that information we could decide how many dwellings to sample within each stratum. That, one hopes, is an improvement over selecting a simple random sample.

Besides the sample mean and total, ratios and proportions are frequently of interest. In some cases the ratio itself is of interest (the bushels of wheat per acre). For the i-th farm raising wheat we get bushels of wheat y_i and acres x_i and take as the *ratio estimate* $\Sigma y/\Sigma x = \bar{y}/\bar{x}$. In other cases we may be interested in the proportion of Republicans who will vote for X. Those proportions or ratios are sometimes used to estimate totals: How many Republican *votes* will X get in this election? For that purpose we multiply the *ratio-estimates* (proportion voting for X) by the expected numbers of Republican voters. It may be that the linear relationship between y and x does not pass through the origin. We then fit the relationship and use the *linear regression estimate* in the way described for the ratio-estimate.

We return now to some types of nonrandom sampling plans. In *quota sampling* the field worker is asked, for example, to go into a specific area and interview 5 white persons, 3 blacks, and 1 Hispanic aged 20 to 24. The interviewer thus has a quota that he fills in a nonrandom fashion. It is one type of *convenience sampling* (using the elements that happen to be "handy"), but it is also *purposive* or *judgmental sampling*, in which the sampler uses his judgment to select the sample. Nothing can be said about the statistical properties of such a sample; it is used when time is of the essence or when no frame is available. Instructions are usually given to help avoid some serious biases.

When no frame exists for sampling a population (of the deaf, say) a device known as *snowball sampling* is sometimes used. The sampler finds 1 deaf person and asks him for the names of other deaf persons he knows. From those contacts, he finds others, etc. Again, nothing can be said about the statistical properties of this nonrandom sample.

A source of very real concern to the sampler is *nonresponse*, the failure to measure some of the units in the sample. The major reasons for nonresponse are the following: (1) *noncoverage* (failure of interviewer to visit the designated home), (2) *not-at-home* (the respondent is not available), (3) *unable to answer* (respondent does not have the information), and (4) *hard-core* (refuses to give the information). If the reason for the nonresponse is related to the information requested, a serious *bias* may result. There are other types of bias related to nonresponse. Well-known types are covered in the section on Epidemiology. Connected with every survey is the job of *data editing*, in which one checks the data to see whether they are in the proper range, complete, etc. Missing data are handled in a variety of ways, including: (1) the use of regression to predict the missing values, (2) the use of the value that minimizes the error sum of squares, and (3) the *hot deck method* (substituting the value of the last observation that appeared in the category of interest). Substituting a value for the missing one is called *imputation*.

The survey *instrument* (questionnaire) is administered by means of a *mail survey, a telephone survey,* or *a personal interview.* For the nonrespondents or not-at-homes, a *call-back* or *follow-up* is frequently made. To compensate for unlisted numbers in a telephone survey, *random-digit-dialing* is used. That consists of calling phone numbers at random within the 3-digit exchange number appropriate for the area. Before an instrument is used, it usually undergoes testing on a small scale, called a *pretest* or *pilot test*. The purpose of pretesting is to check for ambiguity and inconsistency in the questions and unanticipated problems in administering the instrument.

References

A very readable nonmathematical book is Sudman, S. 1976. *Applied Sampling*. New York: Academic Press. The classic reference is Cochran, W. G. 1977. *Sampling Techniques*. 3rd ed. New York: Wiley and Sons. A good treatment is also given in Kish, L. 1965. *Survey Sampling*. New York: Wiley and Sons.

Index